ULTRA HIGH ENERGY COSMIC RAYS: A NEW TOOL FOR ASTROPHYSICS RESEARCH

SPACE SCIENCE, EXPLORATION AND POLICIES

Exploring the Final Frontier:
Issues, Plans and Funding for NASA
Dillon S. Maguire (Editor)
2010. 978-1-60876-080-0

Sustaining the Global Positioning System
Earl M. Peabody (Editor)
2010. 978-1-60741-006-5

Analytical Methods for the Formation of
Dark Matter Haloes in the Universe
Nicos Hiotelis (Author)
2010. 978-1-60876-473-0

Dark Energy: Theories, Developments, and Implications
Karl Lefebvre and Raoul Garcia (Editors)
2010. 978-1-61668-271-2

Pulsars: Theory, Categories and Applications
Alexander D. Morozov (Editor)
2010. 978-1-61668-919-3

Space Material Sciences
A.I. Feonychev (Author)
2010. 978-1-61668-236-1

Ultra High Energy Cosmic Rays:
A New Tool for Astrophysics Research
Anna Uryson (Author)
2010. 978-1-61668-847-9

Commercial Space Transportation
Jocelyn S. Gunther (Editor)
2010. 978-1-61668-707-6

Future of U.S. Human Spaceflight: Background and Issues
Derek A. Warren and Bridget D. Conway (Editors)
2010. 978-1-61668-774-8

NASA's Future and it's Pursuits
Sophia C. Correno (Editor)
2010. 978-1-61668-999-5

Space Science, Exploration and Policies

Ultra High Energy Cosmic Rays: A New Tool for Astrophysics Research

Anna Uryson

Nova Science Publishers, Inc.

New York

For permission to use material from this book please contact us:
Telephone 631-231-7269; Fax 631-231-8175
Web Site: http://www.novapublishers.com

NOTICE TO THE READER

The Publisher has taken reasonable care in the preparation of this book, but makes no expressed or implied warranty of any kind and assumes no responsibility for any errors or omissions. No liability is assumed for incidental or consequential damages in connection with or arising out of information contained in this book. The Publisher shall not be liable for any special, consequential, or exemplary damages resulting, in whole or in part, from the readers' use of, or reliance upon, this material.

Independent verification should be sought for any data, advice or recommendations contained in this book. In addition, no responsibility is assumed by the publisher for any injury and/or damage to persons or property arising from any methods, products, instructions, ideas or otherwise contained in this publication.

This publication is designed to provide accurate and authoritative information with regard to the subject matter covered herein. It is sold with the clear understanding that the Publisher is not engaged in rendering legal or any other professional services. If legal or any other expert assistance is required, the services of a competent person should be sought. FROM A DECLARATION OF PARTICIPANTS JOINTLY ADOPTED BY A COMMITTEE OF THE AMERICAN BAR ASSOCIATION AND A COMMITTEE OF PUBLISHERS.

Library of Congress Cataloging-in-Publication Data

Uryson, Anna.
 Ultra high energy cosmic rays : a new tool for astrophysics research / Anna Uryson.
     p. cm.
 Includes index.
 ISBN 978-1-61668-847-9 (softcover)
 1. Cosmic rays. I. Title.
 QC485.U79 2009
 539.7'223--dc22
                          2010013741

Published by Nova Science Publishers, Inc. ✚ *New York*

CONTENTS

PREFACE

The subject of this book is one of the hot topics in modern physics: the origin of ultra high energy cosmic rays. This subject relates to various aspects of cosmic rays physics and astrophysics. Consequently, the chapters of the book describe astrophysical and cosmic ray issues which are essential for understanding the subject. Topics include a history of cosmic ray research, specifics of detection of cosmic rays at different energy bands, a list of detector arrays for cosmic ray investigation at ultra high energies, GZK-effect, description of the Universe and of some astrophysical objects, particularly of active galactic nuclei.

In this book author also presents original results related to the subject. According to the author sources of cosmic rays in this energy band are active galactic nuclei. Hence author substantiates that cosmic ray data at ultra high energies provide information about active galactic nuclei, namely: particle acceleration conditions, power emitting in cosmic rays, and variability. In addition this book discusses investigation of jets in active galactic nuclei using cosmic ray energy spectra and element composition. Finally, author evaluates the investigation of extragalactic radio background using cosmic ray data.

1. INTRODUCTION

Cosmic rays were discovered more than one hundred years ago, in the beginning of the XX century. Since then cosmic ray investigation has formed a special field of physics, which is called *cosmic ray physics*.

What are *cosmic rays*? Where are they coming from? What are they sources? Cosmic rays are elementary particles and atomic fragments of extraterrestrial origin that bombard detectors with velocities close to the speed of light. Most energetic of them have energies higher than 10^{20} eV while the energies of least energetic are ~11 orders of magnitude smaller. At present we know that particles at different energy bands originate from different sources. Cosmic rays with energies less than 10^9 eV originate from the Sun. Particles at higher energies (up to 10^{18} -10^{19} eV) are born in star processes in our Galaxy. Cosmic rays with energies higher than $4 \cdot 10^{19}$ eV seems to originate outside our Galaxy. Cosmic rays at this energy band ($E \gtrsim 4 \cdot 10^{19}$ eV) are called *ultra high energy cosmic rays*. Depending on the cosmic ray energies different methods and techniques are used to register particles.

This book discusses specific problems related to research in the area of physics of ultra high energy cosmic rays. Studies of cosmic rays at these energies introduce a set of additional challenges. First, extremely low flux: $J \approx 1$ particle/(100 km^2·year). (Imagine a square detector 100x100 km registering a single particle during one whole year of operation). Second, energies of particles can be determined only indirectly with accuracy of 10-30 percent. Another problem is to determine the directions of particles arrivals on the detectors. Finally, the type of the registered particle (proton vs. atomic nucleus vs. gamma quantum, etc.) should be established.

1.1. THE CONTENT OF THE BOOK

Most Challenging Problems in Ultra High Energy Cosmic Ray Physics

What are the most challenging problems in cosmic rays physics at ultra high energies?

Before discussing problems of ultra high energy cosmic rays the book provides background information about cosmic rays with rich in discoveries history of cosmic ray physics that is presented in **Section 2.** This section also includes a brief description of methods of cosmic ray registration and a list of giant ground detector arrays.

Returning to the problem of identification of ultra high energy cosmic ray sources author believes that cosmic ray sources are located outside our Galaxy. Why do we think so? The reason is that apparently there are no objects in our Galaxy where particles can gain so high energies. Hence the conclusion comes that cosmic rays arrive from extragalactic space. Where are they originated? This is the question which has no sure answer yet.

In my opinion astrophysical objects emit ultra high energy cosmic rays. What are these objects? This question is connected with astrophysics, and so **Section 2** also provides a brief description of main astrophysical concepts touched upon in this book.

Section 3 introduces different models of origin of cosmic rays. Using these models we can make attempt to identify cosmic ray sources provided that we know directions where particles came from. Since most of the cosmic rays are charged particles they are inevitably deviated by magnetic fields both in extragalactic space and in our Galaxy. Luckily extragalactic magnetic fields are apparently weak and they are unable to deflect cosmic ray significantly. In the Galaxy, the field structure allows particles to propagate without significant deviation if they arrive neither along the disk of the Galaxy nor perpendicular to the galactic regular magnetic field. In summary both kinds of magnetic fields deflect particles weakly and that makes it possible to identify cosmic ray sources in principle.

Method of cosmic ray identification and some important results are presented in **Section 4**.

The second problem is how particles are accelerated to ultra high energies ($E \gtrsim 4 \cdot 10^{19}$ eV) within the identified objects. Identifying sources, active galactic nuclei of two types appear to be possible cosmic ray sources: low-luminosity Seyfert nuclei and Blue Lacertae objects (BL Lac's). Author proposes a model for cosmic ray acceleration to ultra high energies in low-luminosity Seyfert nuclei, which is presented in **Section 5**.

In **Section 6,** another acceleration model is discussed, which is suitable for Bl Lac's.

In **Section 7,** cosmic ray propagation from extragalactic sources to Earth is discussed comparing cosmic ray data available and results got in the model. Author demonstrates that cosmic ray data provide information about cosmic ray sources: particle acceleration conditions, luminosity in cosmic ray, variability, and in addition parameters of active galactic nucleus jets. Author concludes that cosmic ray at ultra high energies can be used as a tool for investigation of active galactic nuclei.

In **Section 8** author discusses clusters of cosmic ray particles and their origin.

Section 9 provides discussion on cosmic ray propagation in the extragalactic space, from sources to our Galaxy. The well-known GZK-effect decreases particle energies throughout their propagation, caused by cosmic ray interactions with relic emission in the Universe. In addition particle interactions with radio background emission reduce cosmic ray energies as well, though to a lesser degree. Author proposes using GZK-effect with radio photons to examine the intensity of radio background emission in the extragalactic space. At present this study is not completed. In some of energy bands only theoretical values of the intensity are got, but there are no measurements. The proposal and some results are presented in **Section 9**.

Finally, **Section 10** summarizes all ideas and results and provides conclusion.

2. SOME ASPECTS OF COSMIC RAY PHYSICS AND ASTROPHYSICS

2.1. BRIEF OVERVIEW OF COSMIC RAY PHYSICS

2.1.1. Historical Resume

Credit for the discovery of the cosmic rays belongs to Englishman Charles Wilson famous for inventing cloud (Wilson) chamber. In 1900 Wilson conducted experiments with gas in an enclosed vessel and observed the gas to become ionized. To explain ionization Wilson suggested that it was radiation of extraterrestrial origin that penetrated vessels and ionized gas. Later the radiation was named cosmic rays.

However, such experiments were not conclusive enough for this hypothesis to be accepted. Strictly speaking, cosmic rays have been discovered by Austrian physicist Victor Hess who used ionization chambers which were operated on balloons, in 1911–1912. German physicist Kohlherster confirmed Hess's data in 1913–1914.

Both Wilson and Hess are Nobel Prize laureates in Physics, in 1927 and 1936 respectively.

In 1927, Russian physicist D.V. Skobeltsyn demonstrated that the cosmic radiation consists of charged particles. Later it was discovered that it actually is a very small admixture of gamma quanta and neutrinos.

In the 1930-40s the cosmic ray interactions with matter were studied intensively, along with the generation and absorption of secondary particles in the atmosphere. In 1932 and 1936 American physicist Carl Anderson discovered in cosmic rays new elementary particles, a positron and a muon. He is a Nobel Prize laureate in Physics in 1936. Another elementary particle, a pion, was discovered in cosmic rays by C. F. Powell in 1947.

These investigations were performed with the technique widely used in cosmic ray research: balloon borne telescopes composed of Geiger–Muller counters, Wilson chambers, and nuclear photographic emulsions.

By measuring tracks of atomic nuclei in photographic emulsions it is also possible to determine the nuclear composition of cosmic rays. Using this method elements from hydrogen to iron were discovered in cosmic rays in 1948.

In 1961, positrons, about 1 per cent, were observed in cosmic rays in stratosphere measurements.

To conclude, cosmic rays are mostly charged particles with a very small admixture of gamma quanta and neutrinos. Fraction of positrons is only about 1 per cent. At energies $E>2.5$ Gev most (~90 per cent) of the cosmic rays particles are protons. Helium nuclei constitute about 7 per cent, and fraction of other nuclei is less than 1 per cent. At ultra high energies $E>4\cdot10^{19}$ eV, cosmic ray element composition is not clear yet.

2.1.2. The Cosmic Ray Energy Spectrum

According to various authors, the cosmic rays energy spectrum is described by a power law: $J(E) \propto E^{\chi}$, with the exponent $\chi\approx3$. This is valid within an energy range from ~10^{11} to 10^{20} eV.

The value of the exponent varies slightly at the energies about $3\cdot10^{15}$, $6\cdot10^{17}$ and 10^{19} eV. Using logarithmic scale the cosmic ray energy spectrum is a broken line which slope changes at the energies abovementioned. The spectrum is shown in figure 1. In a way it's shape reminds that of the human leg and because of this we say that the spectrum has the "knee" (at approximately $3\cdot10^{15}$, $6\cdot10^{17}$ eV) and the "ankle" (at about 10^{19} eV).

What are reasons for slope changes? Here is the common explanation.

Cosmic rays with energies of ~ $(10^{11}-10^{19})$ eV are originated in our Galaxy. The galactic magnetic field confines particles and prevent them from escaping the Galaxy. However the confinement becomes less effective as the particle energy increases. As a result the particle flux reduces at energies above $3\cdot10^{15}$ eV. That makes the cosmic ray spectrum harder and the "knee" is formed. Additionally, cosmic rays in the region of the "knee" may be emitted by sources having harder energy spectra.

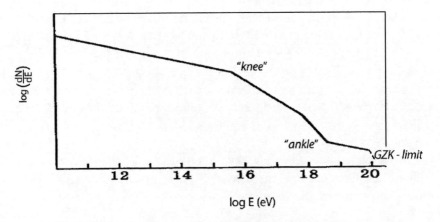

Figure 1. The schematic view of the cosmic ray energy spectrum.

The origin of the "ankle" can be explained in the following way. The cosmic ray flux at energies above the "knee" contains two components, one of which is of the galactic origin (galactic particles) and the second component is of the extragalactic nature. The concentration of the extragalactic particle flux is considerably lower than that of the galactic one. However the galactic cosmic ray flux decreases as the particle energy increases; as a result at energies above $\sim10^{19}$ eV extragalactic cosmic rays become more intensive compared to galactic rays. Hence the particle flux at energies of 10^{19} eV and higher consists mainly of extragalactic cosmic rays that have harder spectrum and this is how the "ankle" is produced. Some details of the cosmic ray spectrum at ultra high energies will be discussed in Section 7.

2.1.3. Extensive Air Showers

At the energies above 10^{14} eV the flux of cosmic ray particles is so low that the use of the foregoing methods becomes inefficient. The cosmic ray flux with energies $E \geq 10^{15}$ eV is $J \approx 3 \cdot 10^{-6}$ $(m^{-2} \cdot s^{-1} \cdot ster^{-1})$. For illustration, let us regard photographic emulsion applicability. Usually photographic emulsions are gathered in stacks of $0.01 m^2$ areas and of π-solid angle. Only one particle at $E=10^{15}$ eV per 120 days falls on such emulsion stack. At the energy of 10^{16} eV one particle per 19 operating years falls on our stack. How can one detect cosmic rays at these energies? In this energy band cosmic ray particles can be registered in another way. The point is that in the atmosphere cosmic particles produce electron-nuclear cascades.

When incident on atmosphere, cosmic ray particles interact with atomic nuclei of air molecules, mainly nitrogen and oxygen. Result of this interaction – secondary pions both charged and neutral that are unstable. The lifetime of charged pions is about $2.6 \cdot 10^{-8}$ seconds, and that of neutral pions is approximately 10^{-17} seconds. However at high energies, the decay length of a secondary charge pion is long enough (due to Lorentz transform), and it interacts with other nuclei in atmosphere. As a result, a nuclear cascade is formed.

Pions with lower energies have short decay lengths and they decay without any interaction. Pion decays give rise to the electron–photon and muon components of the cascade:

$$\pi^0 \to 2\gamma \to 2(e^+ + e^-),$$ (2.1)

$$\pi^+ \to \mu^+ + \upsilon,$$ (2.2)

$$\pi^- \to \mu^- + \upsilon.$$ (2.2)

Particles generated in a cascade are called secondaries. In atmosphere secondaries interact with atoms of air molecules and produce a shower of particles. It is called *extensive air shower*. Air showers are somewhat similar to a brief fall of rain. However air showers are extremely short and extremely wide extended. Their duration is of nanoseconds, and the spread is of several square kilometers.

At every moment of time, an external air shower produced by a cosmic ray particle represents a thin expanding disk of particles moving through the atmosphere in the direction of the primary particle. All particle velocities are close to the speed of light. In the shower all particles fall on detectors almost simultaneously. We call these particles coherent.

Any external air shower has a center of symmetry. The line connecting the shower symmetry centers at different levels in the atmosphere is referred to as the shower axis. The shower axis is the line of the primary particle arrival direction.

What are sizes of external air shower disks? The disk thickness is small, typically about 1.5 meters. The disk radius depends on the number of particles in an external air shower which in turn depends on the cosmic particle energy.

To illustrate, a primary particle with energy of 10^{14} eV, the shower disk size amounts to several hundred meters and contains a huge number of particles.

That allows to use fairly small area detectors distributed over a large field to detect external air showers. For example, an array with a square area of 1 km^2 can record about 100 showers at energies about 10^{18} eV per year.

Can we determine arrival directions of cosmic particles? The answer is "Yes", by measuring the shower axis inclination. Particles detectors are located horizontally. An external air shower having the axis perpendicular to the level, the disk of particles is oriented horizontally and all particles reach detectors simultaneously. In an inclined shower, particles reach detectors with time lags depending on the inclination angle. These lags are used to determine the cosmic ray arrival directions.

With multiple productions of secondary particles taking place in the cascade, secondaries consume almost 90 per cent of the primary particle energy. Ground arrays detect external air showers and determine primary particle energies through energy of secondaries. In arrays, separate detectors are located over large areas of ~10-1000 square kilometers, and because of this they are called giant ground arrays. Typically particle energies are determined with accuracy of 10-30 per cent, and accuracy of the particle arrival direction is about 1^0 (for most of arrays it is about 3^0 and higher).

2.1.4. Registration of Ultra High Energy Cosmic Ray Particles

The flux of ultra high energy cosmic ray particles is extremely low, about 1 particle at the energy $E \approx 10^{20}$ eV per year per 100 km^2. Therefore, giant arrays with area up to tens or even hundreds of square kilometers are required to record ultra high energy particles.

Here is information about ultra high energy cosmic ray arrays:

- Volcano Ranch, USA, operated in 1959–1963, 8 km^2;
- Haverah Park, England, operated since 1967, now inactive, 12 km^2;
- Sydney University Giant Array, (SUGAR), Australia, 1974–1982, 60 km^2;
- Yakutsk, Russia, operated during 1974–1996, 18 km^2 (10 km^2 since 1996);
- Akeno, Japan, operated from 1979 to 1989, 1 km^2;
- Akeno Giant Shower Array (AGASA), Japan, 1990-2004, 100 km^2;

- Fly's Eye, USA, operated during 1981-1992, two detectors at distance of 3.4 km.
- HiRes array (High Resolution, USA), operated in 1997-2006, two groups of detectors at distance of 12.5 km.

Currently measurements are performed on the giant array:

- Pierre Auger observatory, two giant 3000-km^{2}arrays in the Northern and Southern hemispheres.

The ultra high energy cosmic ray statistics collected at different arrays is the following.

- Yakutsk array: 13 showers (including 1 shower at $E > 10^{20}$ eV),
- AGASA: 58 showers (including 8 showers at $E > 10^{20}$ eV),
- HiRes array: 61 showers (including 4 showers at $E > 10^{20}$ eV),
- Pierre Auger observatory: 67 showers (including 2 showers at $E > 10^{20}$ eV).

Typically errors of the inclination angle measurements are in the range from 3^0 to 10^0 (AGASA detector - 1.5-3^0). The typical value for energy resolution for cosmic particles is $\Delta E/E \approx 0.3$. For the Pierre Auger observatory, the angular resolution is about 1.5^0, and the energy resolution is about 0.1.

In addition to giant ground-based arrays, it is possible to observe ultra high energy cosmic rays using satellite-borne detectors, which register the radio signals generated by external air showers. This method allows conducting experiments with huge spreads, almost as big as Earth's atmosphere.

2.2. INTRODUCTION TO ASTROPHYSICS

It is generally accepted that ultra high energy cosmic ray sources are located outside our Galaxy. Consequently one of the main problems of cosmic ray physics is the problem of identifying these sources. Cosmic particles travel long ways through extragalactic and galactic space before reaching an array of detectors on Earth. So next we need to know what physical conditions exist during this journey.

Further discussion requires review of some topics in astrophysics that follows.

The following review relates to cosmic objects that are assumed to produce ultra high energy cosmic particles, conditions under which these particles propagate, and processes they are involved in the extragalactic space.

2.2.1. The Universe

The entire world surrounding our Earth is called the Universe. The Universe has no boundaries, it is infinite and it continuously evolves. This is what we call non-stationary Universe (*non stationarity*). The Universe now is different from the Universe in the past. Physicists reconstruct the history of the Universe using data obtained from astronomical observations as well as results of experiments in laboratories.

The most important feature of the Universe is that it is permanently inflating. This is proved by direct astronomical observations: average distances between galaxies and clusters of galaxies continuously increase. In modern physics it is commonly accepted that before inflation the Universe existed in the form of extremely small volume, in which the matter had extremely high density and temperatures. Conditions of this stage cannot be reproduced in modern lab experiments. This initial hot stage - the epoch when neither stars nor galaxies existed - ended about 12-15 billions years ago and the Universe started to inflate and subsequently cool down. This period is considered as "the moment of birth" of the modern Universe. Theory describing birth and evolution of the Universe is called the Big Bang theory. "Big Bang" is not analogous to any ordinary explosion because it describes rapid expansion of the space which carries matter within itself.

The Big Bang produced powerful radiation called *cosmic microwave background*. This emission still exists in the Universe and since it exists from the very beginning of the Universe sometimes is called *relic*. Its residual temperature is $T \approx 2.7$ K and density is $n \approx 400$ photon cm^{-3}. Cosmic microwave background has two basic features both related to the start of formation of objects in the young uniform Universe. These are angular isotropy and blackbody spectrum. The angular anisotropy is very low (about 10^{-5} of the average intensity) meaning that it almost equal over the whole sky. Studying angular anisotropy allows us to make more accurate determination of certain parameters of the Universe during inflation.

Russian-born American physicist G. Gamov was the first who understood that the emission originated in the very young Universe still exists. The cosmic microwave background was experimentally discovered in 1961 in the USA by A. Penzias and R. Wilson (Nobel Prize laureates in Physics, 1978).

In the Universe the observed matter is formed by stars, gas and dust, plasma, emission at different energy bands, weak magnetic fields, and cosmic rays. However astronomic observations also provide evidence of existence of a non visible matter. This matter is called *dark matter*, and its mass is called *hidden mass*. The hidden mass may amount to 90 per cent of the mass of the visible matter. Many interesting hypotheses exist about the nature of dark matter, but yet it is not determined.

Another important assumption in cosmic physics is that the universe space itself contains energy. It is the main characteristic of the vacuum of the Cosmos. This energy of the space (called *dark energy*) contains about 70 per cent of the total energy of the Universe. At present the dark energy hypothesis is tested in astronomic observations and some assumptions have already been proven.

What is the future of the Universe? Will it inflate infinitely? Or will the inflation stop and the collapse of the Universe occur? The future of the Universe depends on the space, on its dark energy. Some observational evidence exists that the rate of the Universe inflation increases. One assumption believes that the velocity is increasing because the dark energy acts as antigravitation. If this is correct than the inflation will never stop. An alternative assumption predicts some end of inflation and consequent compression and even collapse of the Universe.

The Universe arrangement and its history were discovered in first decades of XX century. Einstein (Nobel Prize laureate in Physics, 1921), who worked then in Germany, Friedmann of Russia, and Lemaitr of Belgium are credited with major contributions.

2.2.2. Units of Measurement

Distances between objects in the Universe are huge and that makes common units of measurements like kilometers or miles inconvenient for its description. For example, smallest distances in the Universe are those within the Solar system. The distance between Earth and the Sun is about 149.6 millions kilometers. Even in this case the special unit, *astronomical unit*, is introduced for simple and convenient description. Distance equal to one astronomical unit is the distance between Earth and the Sun. However even this unit is still too small to measure distances in the Universe.

Distances beyond Solar system are measured in *light years* and *parsecs*. *Light year* is the distance that light travels during a year moving with the velocity of 300 000 km/s. Light year equals 9 461 billions kilometers. Another unit, *parsec* (pc), is the distance equal to 3.26 light years or 30.857 billions kilometers. The word "parsec" and the number "3.26" seem to be somewhat strange. The term "parsec" is a combination of words "parallax" and "(angular) second". For explanation see figure 2. It illustrates the concept of parallax and the origin of number "3.26".

Units longer than parsec are also widely used: kiloparsec (kpc) equals to thousand parsecs and Megaparsec (Mpc) - 1 million parsecs. The size of the Universe is about several thousands of Megaparsecs.

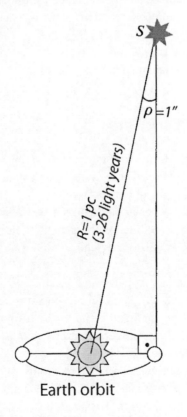

Figure 2. What *parsec* is.

2.2.3. Celestial Coordinates

The visible position of an object in the sky can be defined in various coordinate systems. In this book we use two systems. First system is the *equatorial coordinates*: *right ascension (RA)* α and *declination (Dec)* δ. The second system is the *galactic coordinates*: *galactic latitude b* and *galactic longitude l*. The formal definitions of these coordinates and formulae for equatorial-to-galactic coordinate conversion can be found in astronomical handbooks.

Nonstrict description of these systems is following.

RA is the position of an object measured in the west or east direction on the sky. Dec is the position of an object measured in the north or south direction on the sky. RA is similar to longitude counted from the Greenwich meridian, and Dec is similar to latitude counted from Earth equator.

Galactic latitude is counted from the galactic disc and galactic longitude is measured from the direction to the galactic centre.

It is easy to converse coordinates of an object from one system to another by means of simple formulae of spherical geometry.

2.2.4. Galaxies and Their Types

Stars in the Universe form giant systems of various shapes and sizes, that emit electromagnetic radiation. These systems are called galaxies. Galaxies are similar to star "islands" in the universe space.

In addition to stars, galaxies also contain considerable amounts of dark matter, gas, and dust; they are filled with magnetic fields and cosmic rays. The total energy radiated by a galaxy per unit of time (*emitting power*) is called *luminosity*.

Most galaxies combine in pairs and groups, which in turn enter into clusters and super clusters. A typical group contains several bright galaxies, while a cluster consists of hundreds or thousands of galaxies. There are also separate bright galaxies in the Universe but their number is less than per cent of the total number of galaxies.

The distances between the galaxies in a group range from 100 kpc to 1 Mpc. The distances between separate galaxies, galactic pairs, and groups are bigger - ranging from 1 to 100 Mpc. Distances between clusters range from 10 to 250 Mpc.

It is convenient to consider galaxies as objects consisting of several separate components. The main components are *star disc*, *gas-dusty disc*, and *spheroid star component*. The star component consists of inner part, called *bulge* and of outer *star halo*. In galaxies there is a compact central part which is the brightest and the most dense. It is called *galactic nucleus*. Classification of galaxies is based on comparative sizes of main components or brightness.

Galaxies which discs have weak contrast with the surrounding matter or even cannot be detected are called *elliptical* (notated as E). Elliptical galaxies are ball- or ellipsoid-shaped. This type of galaxies represents about 25 per cent of all galaxies.

All other galaxies are called *disc galaxies*. Disc galaxies are subdivided into the three morphological types: *lenticular galaxies* (notated as SO), *spiral* (S), and *irregular galaxies* (Ir). Their percentages are of ~20, ~25, and ~5 per cent, respectively.

In any spiral galaxy we can observe *spiral arms* where the major part of the interstellar gas is located. Since new stars are born in the gas regions almost all young stars are concentrated inside the arms. Because of this arms are visible. Spiral galaxies are subdivided into *normal* ones (the arms begin immediately from their centers) and *barred-spiral* (SB) ones. The barred-spiral galaxy contains a bright *bar* passing through its nucleus and extending far beyond it. The arms begin at the bar. Spiral galaxies are classified by the relative sizes of their nuclei and disks.

Lenticular galaxies are intermediate type between elliptical and spiral galaxies. Irregular galaxies are the galaxies that cannot be related to any of SO or S types. *Peculiar* (notated as p) *galaxies* have individual shapes and also belong to this group.

All galaxies move and sometimes they get close to each other. At closer distances gravitation effects result in shape deformation of both galaxies and sometimes even leads to merging. Such galaxies are called *interacting galaxies*.

2.2.5. Active Galactic Nuclei

2.2.5.1. The Nature of Active Nucleus

Some galaxies contain a very bright compact nucleus which has an extraordinary feature - a tiny region of the nucleus emits enormous amount of energy. Neither stars nor star explosions can generate so much energy. The nuclei of this type usually exist in both elliptical and spiral galaxies. These nuclei are called *active galactic nuclei.*

What is the nature of active nucleus? Here is a brief explanation. There are massive or supermassive *black holes* in many (or even in almost all) galactic nuclei. In galactic nuclei, the black hole masses range from $\sim 10^6$ to $\sim 10^9 M_\odot$ where $M_\odot$ denotes the solar mass.

Black holes are objects that have very strong gravitation field, so strong that no particle and even light is able to get away. This is the reason they are invisible. In the vicinity of supermassive black holes, tidal forces are so strong that they disrupt stares.

Gravitation forces of the black holes attract all kinds of matter moving near them: whole stars, remnants of disrupted stars, gas originated from supernova explosions surrounding the black hole, and stellar wind. This falling material forms flat hot disc around the black hole. It is called *accretion disc.* Thickness of the accretion disc appears to be ~ 1 pc. However, some authors believe that it can be as thin as $\sim 10^{-2}$ pc. Optically these disks can be optically thin or thick. Topics related to the accretion disks are discussed by Antonucci (1993), Bednarek (1993), and Begelman et al. (1984).

Due to gas particles collisions at velocities near the speed of light the gas in the accretion disc is hot. The falling gas energy converses into the energy of nucleus. A galactic nucleus becomes active when amount of falling matter is high enough. Specifically, for the nuclei to be active the mass of the falling matter should be equal to a few solar masses per year.

Since the gas in the disk is hot it is a source of electromagnetic radiation. In optical and roentgen bands the accretion disc emission constitutes the main part of the active nuclei luminosity.

So the nucleus activity is due to the energy generated near the massive black hole in the galactic centre.

The most important sign of nucleus activity is bright and broad emission in the optical band in the spectrum of active nucleus. As a rule the spectrum of the active nucleus contains very bright and broad emission lines of hydrogen, oxygen, nitrogen, as well as some other elements. Measured Doppler shift and broadening of lines correspond to the gas velocities of thousands of kilometers per second. In addition, the nucleus emission is variable.

Active nucleus ejects gas or particle fluxes in opposite directions along the disk axis at velocities close to the speed of light. These fluxes are called *jets.* Jets are relativistic.

2.2.5.2. Classification of Active Nucleus

Active nuclei are subdivided into the following types: *quasars, Seyfert nuclei, Blue Lacertae objects* (BL Lac's), *radio galaxies*, and *weakly active nuclei*. A galaxy surrounding active galactic nucleus is called *host galaxy*.

Quasars are the most energetic objects in the Universe. The quasar total luminosity (in all energy bands including radio, infrared, optical, ultraviolet, and gamma) ranges up to 10^{46}-10^{47} erg/s. Most quasars (about 90 per cent) are sources of intense radio emission. These are *radio loud* quasars. Quasars without intense radio emission are called *radio quiet*. Sometimes both radio loud and radio quiet quasars are called *quasistellar objects* (QSO). Quasars have noticeable brightness variability in all energy bands and usually they have very broad emission lines. Quasars with extremely rapid brightness variability are called *OVV*. OVV stands for *optically violently variable quasars*. Quasars are much brighter (about 100 times) than the host galaxy and that makes it very difficult to observe host galaxies. So far, all observed host galaxies containing quasars are elliptical.

BL Lac objects are active nucleus with very weak emission lines and with large amplitude variability in the optical band. Their host galaxies are usually observable. They are giant elliptical galaxies.

OVV and BL Lac's sometimes are called *blazars*.

Seyfert galaxies are spirals with active nuclei having broad emission lines. Total luminosity of Seyfert galaxies is in the range of 10^{39}-10^{45} erg/s. The Seyfert nucleus brightness and that of the host galaxy are of the same order of magnitude and these galaxies are observable.

Radio galaxies are characterized by intense radio emission. It is several thousand times stronger than that of our Galaxy. Most of radio galaxies are elliptical giants. Commonly they have jets ejected from the galactic center. These jets are sources of very powerful radio emission. The total size of the radio source including jets amounts to tens or hundreds of kiloparsecs. The total luminosity of a radio galaxy ranges to 10^{45} erg/s.

Weakly active nuclei are nuclei having features of weak activity. They are common in most of galaxies. Our Galaxy also has the nucleus of this type. Compared to active galactic nuclei the luminosity of our Galaxy in the optical band is about 10^{43} erg/s and in the radio band it is about $3 \cdot 10^{38}$ erg/s. Compare it to the luminosity of the Sun that is $4 \cdot 10^{33}$ erg/s.

In the Universe, the number of galaxies with active galactic nuclei constitutes several per cent of the total number of high luminosity galaxies. The volume concentration of active galactic nuclei is the following. The quasar volume concentration is $10^{-7} - 10^{-9}$ (Mpc)$^{-3}$, that of Seyferts is $10^{-4} -$

10^{-5} (Mpc)$^{-3}$, the concentration of radio galaxies is $10^{-6} - 10^{-7}$ (Mpc)$^{-3}$. The volume concentration of normal galaxies is of the order of 10^{-2} (Mpc)$^{-3}$.

The galaxies with active galactic nuclei and the radio galaxies are listed in catalogues of active nuclei and radio sources, respectively. These catalogues are based on observational data, they contents are renewing and expanding constantly. The corresponding references are given later in the book. Some problems of active galactic nucleus physics are discussed in Section 3.2.1.

2.2.6. Hubble's Law

Astronomical observations show that all distant galaxies move away. Therefore distances between galaxies increase. This phenomenon was discovered by American astronomer E. Hubble in 1929. (The telescope onboard of the space observatory is named "Hubble telescope" in his honor.)

How did Hubble come to this conclusion? He studied light emitted by the stars in the distant galaxies. He observed that spectral lines of stars do not have exactly the same wavelengths as the lines observed in the laboratory, but they were systematically shifted to longer wavelengths, towards the red end of the spectrum. Hubble interpretation is based on the Doppler effect: galaxies are moving away from our own Galaxy, that is why their spectral lines are shifted to the red end of the spectrum.

Why are they moving? Remember, in the inflating Universe space expands and that makes galaxies to move. In a way it reminds of an inflating balloon: distances between points on the balloon surface increase with increasing inflation .

Because of that we observe a spectral line from a galactic source not at the emission wavelength λ_0 but at the wavelength $\lambda > \lambda_0$. The relative change in the wavelength is called *red shift*. It is denoted by z: $z = (\lambda - \lambda_0)/\lambda_0$.

Hubble also discovered the relationship between the speed of a distant galaxy and its remoteness: the speed increases with galactic remoteness. Namely the speed V of a galaxy is proportional to the remoteness D: $V = H \cdot D$, where the coefficient H is called the Hubble constant.

The speed of galaxy is determined from the red shift of it spectrum. The speed of a distant galaxy moving from us is $V = cz$, where c is the speed of light and therefore the distance from a galaxy is

$D = cz/H$, Mpc.   (2.4)

This is *Hubble's law*.

Hubble's law is true for distant galaxies. They are located at distances more than 5-10 Mpc from our Galaxy: $D>(5-10)$ Mpc. These distances are called *cosmological distances*.

Determining the Hubble constant appears to be quite challenging. At present it is determined to be $H=75$ (km·s)/Mpc. Some astronomers assume that the value of H changes with time because the speed of inflation could have another value in the past.

For galaxies with red shift $z \geq 0.5$ relationship between the value of z and the distance depends on the shape of the Universe. Various models of the Universe are accepted in modern astrophysics. The relation between z and D depends on models. One of them is discussed in Section 7.

2.2.7. Our Galaxy Milky Way

Our Galaxy Milky Way is a spiral galaxy with a nucleus which seems to be a weakly active one. The Galaxy itself is a part of the *Local Cluster* the size of which is about 2 Mpc. The Local Cluster is in turn a part of the *Local Supercluster* which is about 30 Mpc in size. Sometimes the Local Supercluster is called *Virgo Supercluster*.

Milky Way is a huge star island containing hundreds of billions of stars. In spite of the enormous number of stars it is a rarefied system: distances between neighbor stars are of several light years. The size of the galactic disk is about 20 kpc. Thus its diameter is more than 10^5 light years. The thickness of the galactic disk is several hundreds of parsecs. The Solar System is located at a distance of about 10 kpc from the galactic center. In addition to stars our Galaxy also contains many objects with low masses (for example, planets).

The galactic interstellar medium is not empty. It contains rarefied gas, dust, magnetic fields and cosmic rays.

Both the galactic disk and the halo contain gas. However it is mainly concentrated in the spiral arms in the disk. Mean gas densities inside the disk and the halo are equal to 1 and 0.01 cm^{-3}, respectively. The main component of the galactic gas is hydrogen that forms *clouds* both atomic and molecular. If there are young stars inside hydrogen clouds their emission ionizes gas and heats it to about 10^4K. The clouds of ionized hydrogen emit light. Molecular clouds are dense and cold compared to clouds of atomic hydrogen; their temperature is only about 10-20 K. Molecular clouds contain also complex molecules, sometimes even organic.

Interstellar gas in the Galaxy is intermixed with dust granules with sizes of ~0.1 micron. Dust granules are composed of carbon, silicon, and gas molecules.

The magnetic field in the Galaxy consists of two components: regular field and random field. The field lines of the regular component are directed along the disk plane. They are approximately parallel to the spiral arms. The mean value of the regular magnetic field is of $3 \cdot 10^{-6}$ G. The random field is of $\sim 10^{-6}$ G, its scale ranges from 30 to 300 pc.

The galactic disk is surrounded by a large gaseous halo with magnetic fields and cosmic rays. It has the same diameter as that of the disk and the thickness which is estimated to be in the range of 3-10 kpc. The magnetic field in the halo is comparable to that in the disk. In the halo a regular magnetic field appears to exist in addition to random component. The direction of the regular field lines is not known yet. It is assumed that the regular field is perpendicular to the disk or directed along it.

The galactic magnetic field governs the propagation of cosmic rays with energies $E < 10^{18}$ eV. In the Galaxy these particles move in a complicated manner due to scattering in the magnetic field. In a way the process of particle propagation reminds the process of diffusion of gas molecules. As a result by the time the cosmic particles reach the Earth they "forget" they own initial directions.

Particles with energies above 10^{17} eV have Larmor radius exceeding the characteristic sizes of the Galaxy (respectively ~5 and ~20 kpc). Because of this they are not confined by the galactic magnetic field and escape the Galaxy.

2.2.8. Two Phenomena as Possible Cosmic Ray Sources

These two phenomena are *gamma-ray bursts* and *topological defects of vacuum*.

Gamma-ray bursts are flashes of gamma rays. They are the most luminous events in the universe. It is generally accepted that bursts are associated with extremely energetic explosions in distant galaxies. Gamma bursts are detected by satellites. Duration of bursts ranges from milliseconds to nearly an hour with typical value of a few seconds. Totani (1998) suggested that gamma bursts can produce cosmic rays due to giant energy release.

Topological defects of vacuum or defects in vacuum structure were created in the universe during the first 10^{-37} s after the Big Bang. After Vilenkin

(1996) the decay of metastable defects could produce cosmic rays with energies above 10^{20} eV.

2.3. COSMIC RAYS IN THE GALAXY AND IN THE EXTRAGALACTIC SPACE

2.3.1. Galactic Cosmic Rays

It is generally accepted that cosmic rays with energies less than $10^{18} - 10^{19}$ eV are originated in the Galaxy. Their sources are *supernova* and *pulsars* - very powerful sources, capable of cosmic ray acceleration. A supernova is the explosion of a giant star. In this explosion the star nucleus collapses and turns into a *neutron star*, while the shell of star gas is stripped. The shell expands and travels away from the star. This shell is called the *supernova remnant*. Some of neutron stars are magnetized.

Pulsars are magnetized neutron stars. In pulsars, the surface magnetic field is of 10^{10}-10^{14} G.

Particles with energies about 10^{17} eV are accelerated in supernova remnants. Cosmic rays at higher energies are accelerated most likely in pulsar magnetospheres. What is the mechanism of their acceleration? In supernova remnants particles are accelerated on *shock fronts*. In pulsars, cosmic rays are accelerated by electrostatic potential or by electromagnetic waves. The physics of acceleration mechanism is discussed in books by Longair (1981) and Berezinsky et al. (1990).

2.3.2. Cosmic Ray Propagation in the Extragalactic Space. GZK-Effect

In addition to the cosmic particles the Universe is filled with cosmic microwave background radiation. It was discovered in 1965. Very soon after that physicists realized the important role this radiation played in propagation of cosmic particles. It turned out that as a result of cosmic ray-radiation interactions in the intergalactic space the former are loosing energy.

Cosmic protons p and nuclei A interact with cosmic microwave photons γ_{rel} via reactions (here γ, N denote a photon and a nucleon produced):

$$p + \gamma_{rel} \rightarrow p + \pi^0; \ \pi^0 \rightarrow 2\gamma, \tag{2.5}$$

$$p + \gamma_{rel} \rightarrow n + \pi^+; \ \pi^+ \rightarrow \mu^+ + \upsilon, \tag{2.6}$$

$$A + \gamma_{rel} \rightarrow A' + kN$$

(nuclear photodisintegration, or nuclear photoeffect), $\qquad$ (2.7)

$$A + \gamma_{rel} \rightarrow A' + kN + m\pi \ \text{(photoproduction of pions).} \tag{2.8}$$

These reactions have a threshold, they occur when the energy of cosmic particle is above it. In the cosmic ray system the threshold energy for the reaction of nuclear photodisintegration is $E_{thr} \approx 10$ MeV and that for reactions with pion production is $E_{thr} \approx 145$ MeV. So cosmic rays with energies E_{thr} and higher interact with cosmic microwave background and loose energy in the extragalactic space. Because of this the number of cosmic particles at ultra high energies decreases compared to the initial value and that leads to deformation of the cosmic ray energy spectrum. This effect was discovered in 1966 by German and Russian physicists K. Greisen, G.T. Zatsepin, and V.A. Kuzmin and has the name of *GZK-effect*. (GZK is the abbreviation formed by taking first letters of their last names.)

These physicists calculated the cosmic ray energy spectrum assuming cosmic ray interactions with cosmic microwave background in the extragalactic space. They discovered an abrupt drop in the spectrum at energies $E > 4 \cdot 10^{19}$ eV and a cutoff at energies of 10^{20} eV as compared to the cosmic ray spectrum calculated using power extrapolation.

The effect of cosmic ray spectrum transformation in the extragalactic space is also called the *GZK-cutoff*.

In reaction 2.5-2.8 the mean free path of a particle is above several tens of Megaparsecs. Therefore cosmic particles that travel shorter distances do not experience significant energy losses and they do not have GZK-cutoff in the spectrum.

Cosmic ray propagation in the extragalactic space and the energy spectrum shape was first analyzed in detail by Stecker (1968), Hillas (1984), Hill and Shramm (1985), and by Berezinsky and Grigorieva (1988). In this book the shape of the spectrum is discussed in Section 7.

Another result of cosmic ray interactions with the cosmic microwave background is electromagnetic cascades originated in the space. This effect was discovered by Hayakawa (1966) and Prilutsky and Rozental (1970).

Author believes that certain characteristics of the background emission as well as that of cosmic ray sources can be derived through study of extragalactic electromagnetic cascades. Some original results of the author are presented in Section 9. Author designates cosmic ray interactions with the cosmic microwave background on the whole as the GZK-effect.

3. BRIEF DESCRIPTION OF THE MODEL ACCEPTED IN THIS BOOK

3.1. THE HISTORY OF THE PROBLEM

As we discussed cosmic particles with energies $E \geq 10^{20}$ eV are generally believed to be of extragalactic origin. However several cosmic ray arrays registered particles which energies seem not to be reduced via the GZK-effect. Why? Various ideas have been suggested to explain this phenomenon.

The first group of ideas is based on the assumption that the cosmic ray spectrum has no GZK-cutoff because particles are produced not far enough from the Earth. A number of hypotheses were developed based on this assumption. Ultra high energy particles can be generated in the galactic halo as a result of decays of superheavy relic particles produced as a result of the Big Bang. Also particles may be generated by topological defects of vacuum (loops, strings, etc.) in various parts of the Universe, including the vicinity of Earth. Finally, ultra high energy particles can originate directly in the terrestrial atmosphere by decays of some hypothetical particles or objects.

The second group of ideas is based on the fact that the Lorentz factor depends on velocity of the particle. In this case when this velocity is close to the speed of light the interaction of cosmic rays with the cosmic microwave background can be suppressed or even forbidden. As a result, cosmic rays passing cosmological distances have energy losses strongly reduced compared to the case of the common Lorentz factor.

These hypotheses explain lack of the GZK-effect in the ultra high energy cosmic rays and some of them predict measurable values of parameters in cosmic ray flux. Inquiring readers can find references on this subject on the Internet.

3.2. THE MODEL

Cosmic ray sources are divided into three groups. First, - astrophysical objects of various types: pulsars, active galactic nuclei, components of powerful radio galaxies, quasars, gamma-bursts, and interacting galaxies. Second - ultra high energy particles are generated by cosmological peculiarities of the space under some conditions. Third - decays of relic particles produced as a result of the Big Bang generate cosmic particles at ultra high energies.

Is it correct to identify directly cosmic ray sources?

Following this classification we can assume that it is correct for the first group: direction of the particle arrival points out to object of its origin or, in other words, shower axes are directed towards particle sources. So searching for cosmic ray sources we should look for sources among objects located in the proximity of the shower axis.

For the second group any objects located in the vicinity of particle arrival directions occur there randomly. Models associated with the third group predict that the cosmic ray flux mainly comes either from the galactic center or from the halo with a very weak flux arriving from the Virgo cluster.

In the book author assumes that ultra high energy cosmic rays are of extragalactic origin and their sources are astrophysical objects.

This model is not new; it was widely discussed in 1960s. Then it was suggested for the first time that active galactic nuclei were sources of cosmic rays. Estimates of total power and volume density of active galactic nuclei made by Berezinsky et al. (1990) show that the model is quite realistic.

In the book author examines for the first time objects located at distances $R \lesssim 50$ Mpc from Earth. Author selects these distances because during propagation cosmic particle energy is decreased not much through GZK-effect as shown by Stecker (1968).

3.3. COSMIC RAY DEFLECTIONS IN THE GALAXY AND IN THE EXTRAGALACTIC SPACE

In general cosmic rays are charged particles and therefore magnetic fields affect their trajectories both in the extragalactic space and in the Galaxy. Unfortunately neither structure nor strength of extragalactic magnetic fields is known in detail now.

For this model author assumes that ultra high energy cosmic ray trajectories are rectilinear and during travel to our Galaxy particles are only weakly deviated: the deflection is not more than $\alpha_0=9^0$. ($\alpha_0=9^0$ is the value of the 3-error box around the particle arrival direction, assuming the accuracy of the shower axis determination is 3^0.)

In principle, extragalactic cosmic rays can be quanta. Shinozaki et al. (2003) studied this problem and revealed that ultra high energy cosmic rays were particles (protons or atomic nuclei or their fragments) rather than quanta.

The simplest way to estimate the field that causes this deflection is as follows. The deflection reaches its maximum when a cosmic proton travels perpendicularly to the field and the field inhomogeneities may be ignored. Under this assumption author got that extragalactic magnetic field was $B<10^{-9}$ G.

***

The trajectory of a charged particle with energy E and charge Z for the simplest case described above is shown in figure 3. We look for sources located in the cone within the axis that coincides with axis of the shower. From the source to the detector the particle moves along an arc L. The arc length is: $L=2\alpha r_B$, where r_B is the Larmor radius, which is equal to (Ginzburg and Syrovatskii, 1964)

$$r_B=E/(300ZB). \tag{3.1}$$

In this formula r_B is in cm, and B is in G. The aperture semiangle is related to the angular measure of the arc L^*: $\alpha= L^*/2$. For $L\approx R$ and $\alpha=\alpha_0$ we get when $\alpha\leq10^0$: $B=2\alpha_0E/(300ZL)$. Assuming $R\lesssim50$ Mpc, $Z=1$, and $E\approx10^{20}$ eV we get $B\leq8.7\cdot10^{-10}$ G. Assuming heavy Fe nucleus ($Z=26$) rather than protons, $B\leq3\cdot10^{-11}$ G.

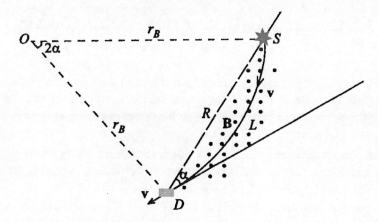

Figure 3. Motion of a particle in the plane perpendicular to the intergalactic magnetic field **B**. *S* is the source, *D* is the detector, *O* is the centre of the Larmor orbit, *L* is the arc along which the particle moves, and α is the angle between the direction toward the source and the shower axis.

***

This value is in agreement with astrophysics estimates. First, the similar value is received from rotation measure of quasars at $z = 2.5$. Second, it agrees with some theoretical limits on possible extragalactic regular relic field. Cosmic magnetic fields are described by Kronberg (1994) and by Wielebinski and Beck (2005).

In addition extragalactic magnetic field seems to have filament structure with regions where the field exists being relatively thin. Because of this particles are only weakly deviated while traveling along short trajectories in the filament regions (Dolag et al. 2004).

In the Galaxy, the particles move in fields $B \sim 10^{-6}$ G. According to Beck et al. (1996) in the disc, the magnetic field in spiral arms is regular and directed along them. In the halo, the regular field is supposed to be perpendicular to the disc. The nonregular field component fluctuates with the reversal length $\delta L \sim 100$ pc, $\delta B \sim 10^{-6}$ G. The deflection of a charged particle in the field with a regular component depends on the initial particle direction and can be small.

In the halo cosmic ray deflections in the random magnetic field also appear to be small. This was shown by Parker (1991).

Based on these remarks on magnetic fields, author proceeds with identification of cosmic ray sources on the celestial sphere using cosmic ray data at ultra high energies.

***

Magnetic field inside galactic clusters is actually much stronger, $B \sim 10^{-6}$-10^{-7} G, as it is discussed in literature. However in the model the field value is of no importance because galactic clusters look like point sources. Why? Cluster sizes range from $L \approx 1$ Mpc to $L \approx 5$-7 Mpc. Hence for clusters located at distances $D \approx 40$ Mpc angular dimensions α_{cl} range from $\alpha_{cl} \approx 1.4^0$ to $\alpha_{cl} \approx 10^0$. (For more distant clusters angular dimensions are even smaller.) Since we are looking for clusters within aperture $2\alpha_0 = 18^0$ any cluster is contained inside this aperture as $2\alpha_0 > \alpha_{cl}$.

***

The mean angular deflection in the random magnetic field was estimated by Cronin (1996).

The mean angular deflection ψ of a cosmic proton due to the walk in the random magnetic field is following (Cronin, 1996):

$$\psi \approx 1.7^0 (d/30 \text{ Mpc})^{1/2} \, (\delta L/1 \text{ Mpc})^{1/2} \, (B/10^{-9} \text{ G})(10^{20}/EeV), \qquad (3.2)$$

where d is the distance to the source, δL is the reversal length of the magnetic field, and B is the mean magnetic field. The particle angular deflection is $\alpha_0 = 1.7^0$ assuming $d \approx 30$ Mpc, $\delta L \approx 1$ Mpc, and $B \approx 10^{-9}$ G. So it appears to be negligible.

4. IDENTIFICATION OF SOURCES OF ULTRA HIGH ENERGY COSMIC RAYS

Assuming small particles deflections in extragalactic magnetic fields author studies astrophysical objects appearing in the region around each shower axis The size of the region is one, two or three error boxes of particle arrival coordinates. Errors in determining object locations are disregarded, as they are in seconds.

Many objects of various kinds can be found within the error-boxes. It is clear that not all of them are cosmic ray sources. What object generates a particle? What objects appearing in the field of search are not related to the particle? Author calculates probabilities P of random coincidence of object and particle coordinates, within error boxes of the latter. According to the law of probability objects with low values of P, $P \sim 10^{-3}$, are located not randomly near the shower axis. Objects having $P \gg 10^{-3}$ occur randomly in the field of search.

4.1. SHOWERS WHICH WERE ANALYZED

Cosmic ray sources were identified in the paper (Uryson, 1996). It appeared that sources were nearby Seyfert nuclei with red shifts $z \leq 0.0092$. It means that they are located at distances less than 40 Mpc assuming Hubble constant to be $H=75$ km/(s Mpc). Then author analyzed 17 Akeno showers with energies $E > 3 \cdot 10^{19}$ eV.

This study was continued in (Uryson, 1999; 2001a, b). In (Uryson, 2001a, b) it was found that BL Lac's are possible cosmic ray sources too. BL Lac's were identified as possible sources also by Tinyakov and Tkachev (2001).

Here author performs the source identification for 63 ultra high energy showers which arrival coordinates were published at the moment, with errors in arrival directions $(\Delta\alpha, \Delta\delta) \lesssim 3^0$ (in equatorial coordinates.) The selected showers and their energies are as follows:

I   58 events with $E > 4 \cdot 10^{19}$ eV (Takeda et al. 1999; Hayashida et al. 2000);

II   4 Yakutsk showers with $E > 4 \cdot 10^{19}$ eV (Afanasiev et al. 1996);

III 1 Haverah Park shower with $E \geq 10^{20}$ eV (Watson, 1995).

Eleven of these showers have energies $E > 10^{20}$ eV. The most energetic shower with energy $E \approx 3 \cdot 10^{20}$ eV registered at the Fly's Eye array (Bird et al. 1995) was not considered because of the bigger coordinate errors: $\alpha = 85.2 \pm 0.4^0$, $\delta = 48.0(+5.2, -6.3)^0$.

The complete list of showers is presented in Table in Appendix.

4.2. ASTROPHYSICAL OBJECTS WHICH WERE EXAMINED

Author considers Seyfert galaxies and BL Lac objects, radio galaxies and roentgen pulsars as possible sources of cosmic rays. The reason is that these objects are generally accepted as possible cosmic ray accelerators. Radio galaxies are one of the most powerful objects among galaxies. Roentgen pulsars are the most powerful of pulsars.

Astrophysical objects are listed in catalogues which are reexamined and renewed in accordance with latest astronomic observations. Author used following catalogues.

- Active galactic nuclei: Lipovetskii et al. (1988), Veron-Cetty and Veron (1993, 1998, 2001, 2003);
- Radio galaxies: Kuhr et al. (1981) and Spinrad et al. (1985);
- Roentgen pulsars: Popov (2000).

4.3. METHOD OF IDENTIFICATION

There are three important issues that should be considered while identifying cosmic ray sources.

First, results of analysis depend on statistics of astrophysical objects and of showers as well. Therefore any analysis would not be complete without this consideration. (This is discussed in Section 4.5.3.)

Second, there are celestial regions with very low number of objects. These are so called *"avoidance zones"* located at low galactic latitudes. Gas and dust of galactic disc obscure objects within these zones and because of that catalogues contain only a few of any objects from these regions.

In addition catalogues of astrophysical objects are not complete - they do not contain all objects of a certain type. Moreover, number of objects in various celestial regions can differs because some zones were observed for a longer time and under better conditions.

With this in mind to decrease the influence of "avoidance zones", author divided showers in groups depending on the arrival galactic latitudes. These groups are:

- 54 showers from latitudes $|b|>11.2^0$,
- 37 showers from $|b|>21.9^0$,
- 27 showers from $|b|>31.7^0$, and
- 63 showers without any selection in Galactic latitude.

Author determines probabilities P of random coincidence of object coordinates with shower axis coordinates within the shower error-boxes. The problem is formulated as follows.

Suppose we register K showers, and N of them have at least one object of the certain type near the axis. Are these objects cosmic ray sources? In other words, are these objects located there by chance, or not?

Author simulates K particles with random celestial coordinates (α, δ), with errors $(\Delta\alpha, \Delta\delta) \approx 3^0$. Simulated showers have arrival directions in the same latitude range as that of the registered showers. After that author determines how many particles (denoted N_{sim}) have at least one object of the same type in the field of search. The number of these particles is within the limits $0 \leq N_{sim} \leq K$. Author repeats this procedure M times ($M=10^5$) and that yields M simulated sets with K random particles. Next author determines how many particles (denoted I_{sim}) in each set have $N=N_{sim}$. The I_{sim} takes two values: 1 if

$N=N_{sim}$ and 0 in other cases. The probability P to find at least one of the non-generating objects in the field of search of N showers among K particles is equal to

$$P= \sum_{i=1}^{M} (I_{sim})_i / M .$$ (4.1)

If all objects of a given type occurred in the field of search randomly (and they are non-generating) probability is $P\sim 1$. For $P\ll 1$ the hypothesis about random occurrence is rejected at the confidence level $1-P$.

4.4. RESULTS

Catalogues of active galactic nuclei contain objects which type is determined surely along with objects which type is not quite clear due to lack of observation. The latter are referred to as possible or probable objects of a certain type. So catalogues contain not only Seyferts and BL Lac's but also objects that belong to one of these types only probably. Because of this author actually calculates probabilities of random occurrence of active galactic nucleus near shower axis both for Seyfert nuclei and possible Seyfert nuclei, as well as for BL Lac's and possible BL Lac's. Values of probabilities are listed in Table 1.

Values of probability P for objects in 1-, 2-, and 3-error boxes around shower axis are presented in (Uryson, 2007). There probabilities are given also for showers arrived from various latitude bands.

Table 1. Values of probabilities *P* of object random occurrences near shower axis and statistics of objects in the band α=0-24 h, δ=-10-90⁰

Correction:

Table 1. Values of probabilities *P* of object random occurrences near shower axis and statistics of objects in the band α=0-24 h, δ=-10-90^0 where showers were registered

Type of objects	Probability P	Statistics of objects
All Seyferts ($z\leq 0.092$)	$P<10^{-3}$	103
Well-defined Seyferts ($z\leq 0.092$)	$P\approx 10^{-3}$	93
All BL Lac's	$P\lesssim 10^{-5}$	438
Well-defined BL Lac's	$P\lesssim 10^{-5}$	292
Radio galaxies	$P\approx(0.01-0.1)$	306
Roentgen pulsars	$P\sim 0.01-0.1$	15

Based on the law of probability author concludes that possible ultra high energy cosmic ray sources are Seyfert nuclei with red shifts $z<0.01$ (i.e. located at distances not more than 40 Mpc) assuming Hubble constant $H=75$km/(Mpc·s)), and BL Lac's. Radio galaxies and roentgen pulsars are excluded from the list of possible cosmic ray sources.

Active galactic nuclei identified as possible cosmic ray sources are listed in Table in the Appendix. Identified Seyfert nuclei have relatively low luminosities and emit comparatively weak fluxes in radio and roentgen bands.

Physicists from Pierre Auger collaboration recently came to the same conclusion that cosmic ray sources are nearby Seyferts with such characteristics (The Pierre Auger Collaboration, 2007a).

4.5. DISCUSSION

Now author will discuss limitations of the method and applicability of the results.

4.5.1. Deflection of Particles

The method of source identification and subsequently obtained results are valid if extragalactic magnetic field is rather weak, $B<<10^{-9}$ G. However magnetic fields inside galactic clusters are much stronger. Still, since any cluster at distances of tens Mpc and higher can be considered as a point source the case when a cosmic ray source is a member of a galactic cluster can be analyzed in the same manner.

Now let us discuss sizes of the region that should be chosen to search for possible sources. Two factors define limits for the size of the region. The first one is errors in particle arrival directions ($\Delta\alpha$, $\Delta\delta$). The second factor is related to particle deflection in the extragalactic magnetic field. The stronger factor determines the choice of the size.

The field $B<<10^{-9}$ G was obtained for particle deviations $\alpha=9^0$ when sources are located at distances $R\leq50$ Mpc from Earth. However, BL Lac's are located at distances of hundreds Mpc. Is it possible that particle deviations in extragalactic fields are small if cosmic ray sources are so distant? This problem was discussed by Dolag et al. (2004). They came to the conclusion that the extragalactic field is localized mainly in relatively thin filaments and therefore cosmic rays deviations are actually small.

4.5.2. One Factor Not Accounted for

Any cosmic ray array observes different celestial regions during nonequal period of time. As a result the number of registered particles is proportional to the duration of observation. Author does not consider this factor in the probabilistic analysis.

4.5.3. Active Galactic Nuclei and Shower Statistics

Active galactic nucleus statistics undoubtedly affects the results. Results were obtained using following statistics for the objects:

- 438 BL Lac's, including 292 objects of surely determined type,
- 103 Seyferts with $z<0.01$, including 93 objects of type determined surely,
- 306 radio galaxies, and
- 15 roentgen pulsars.

Do our results depend on object statistics?

The number of radio galaxies is very close to the number of surely determined BL Lac's. In addition, almost all showers have both radio galaxies and BL Lac's in error boxes around the shower axis. However probabilities for BL Lac's and radio galaxies to occur near shower axes randomly differ considerably.

The number of Seyfert nuclei with $z<0.01$ is about 5 times smaller than the number of BL Lac's. Fewer showers contain Seyfert nuclei in the field of search compared to BL Lac's. However both Seyfert nuclei and BL Lac's have low probabilities of random occurrence near shower axis.

That brings us to the conclusion that Seyfert nuclei and BL Lac's are cosmic ray sources. Also author concludes that radio galaxies do not emit ultra high energy particles. Is it possible to make the same conclusion regarding roentgen pulsars? At present it seems incorrect because their number is ~10-30 times lower than number of other objects under consideration.

Does the number of analyzed showers actually affect the results? The answer is "Yes". Comparing results for 17 showers published in (Uryson, 1996) with that for 63 showers analyzed here it appears that values of probability P are different depending on the shower statistics.

4.5.4 Interacting Galaxies as Possible Cosmic Ray Sources

As it was suggested by Cesarsky (1992) conditions for particle acceleration also exist in the interacting galaxies. However author does not consider interacting galaxies as possible cosmic ray sources. The reason is that most of normal galaxies are also interacting (Wright et al. 1988), so the probability to find interacting galaxies near the shower axis is high whether or not interacting galaxies are cosmic ray sources.

For identification, it is necessary to select interacting galaxies where particle acceleration is effective among a bulk of normal galaxies. Currently available observation data are not sufficient to surely select interacting galaxies.

4.5.5. Dominant and Subdominant Cosmic Ray Sources

Statistical approach described here reveals dominant (most numerous) cosmic ray sources, but it does not exclude some other hypotheses. For example, how we should interpret the fact that some showers have no astrophysical objects in the search area. One of the possible explanations is that catalogues are incomplete and contain not all objects of a certain type. However the same fact of empty error-boxes can be explained keeping in mind that statistical approach points out to dominant, more numerous or more powerful sources, but there are other sources, less effective or less abundant (rare). These minor sources emitting cosmic rays, particles seem to have no dominant sources in their error-boxes.

4.6. CONCLUSION

Using statistical analysis of particle arrival directions sources of cosmic rays are identified. With some reservations mentioned above it is concluded that ultra high energy cosmic ray sources are active galactic nuclei, namely low-luminosity Seyfert nuclei located at red shifts $z<0.01$, and BL Lac's. Results are valid when extragalactic magnetic field outside galactic clusters is weaker than $B<10^{-9}$ G and deviations of particles traveling through the Galaxy are negligible.

So the next question arises: how are particle accelerated in the identified sources? This is discussed in the following section.

5. THE MODEL OF COSMIC RAY ACCELERATION IN SEYFERT NUCLEI. PREDICTIONS OF THE MODEL

5.1. PARTICLE ACCELERATION IN SEYFERT NUCLEI

Different authors suggested various objects as possible sources of cosmic rays. All this objects have an important common feature: they are powerful. However Seyfert nuclei that were identified as cosmic ray sources are low-luminosity, rather than powerful objects. So how are particles accelerated to ultra high energies in these objects? Here a model of cosmic ray acceleration in low-luminosity Seyfert nuclei is discussed. In spite of the fact that different models of cosmic particle acceleration are discussed in literature author suggests one more for the reason that in the sources various conditions can exist and therefore different acceleration processes can be realized.

Starting these studies in the mid 1990s author assumed that in most Seyfert nuclei jets are formed in the vicinity of a central black hole and low-luminosity Seyfert nuclei have relativistic jets of several parsecs in length. At that time there were no experimental observations of these objects and most of the studies were based on some assumptions. By later, jets of this length have been observed in numerous low-luminosity Seyfert nuclei (Xu et al. 1999; Falcke et al. 2000; Thean et al. 2000; Ulvestad and Ho, 2001). However neither the mechanism of jet production nor the jet composition on very small scales is still unknown, as one is not able to distinguish between the various theoretical models using existing experimental observation.

The magnetic field is frozen in the jet material. Observations show that in some jets the field is oriented along them, and there are jets in which the field is directed across the jet (Gabuzda and Cawthorne, 2003). Magnetic fields in jets were investigated theoretically by Istomin and Pariev (1994).

In the model author considers low-luminosity Seyfert nuclei with relativistic jets of ~1-3 parsecs in length and the magnetic fields oriented along these jets.

Jets disturb the surrounding material and as a result cocoons of perturbed jet material are formed around jets. Shock waves are exited in the jet and in the cocoon while they propagate in the vicinity of the central black hole (Blandford and Eichler, 1987). Shocks are exited due to the nonlinear development of instabilities at the jet surface, collisions with clouds, and fluctuations of the jet velocity. Krymskii (1977) revealed that at shocks particles can be accelerated via diffusive shock acceleration. (It is a subclass of Fermi acceleration.)

Author assumes that

1) particles are accelerated in jets at shock fronts with a regular magnetic field via diffusive shock acceleration, and
2) cosmic rays are accelerated to ultra high energies by this mechanism.

However concurrently with acceleration, particles moving in magnetic field also loose energy via synchrotron radiation.

For this model author accepts that the particle can gain maximum energy if synchrotron losses do not exceed half of energy gained during acceleration.

The value of the magnetic field in Seyfert jets is not very well known. Because of this author introduces it as a parameter of the model. Author assumes that the magnetic field is in the range of 5-1000 G, according to estimates derived by Rees (1987), Sikora et al. (1997), and Bednarek (1993).

Next, author assumes that the jet contains some material from the accretion disc, and cosmic ray composition is determined by the disc chemical composition.

Using these assumptions author received the value of magnetic field B_{CR} in which the particle with the charge Z gains maximum energy

$$B_{CR} \approx 50 Z^{-1/3}. \tag{5.1}$$

Nuclei are accelerated to the maximal energy

$$E_{\text{max A}} \approx 6.6 \cdot 10^{20} \, (Z/B)^{1/2} \text{ eV}, \tag{5.2}$$

and protons to

$$E_{\text{max p}} \approx 1.65 \cdot 10^{20} \, B^{-1/2} \text{ eV}. \tag{5.3}$$

***

In parallel shocks, with the jet field parallel to the jet axis particles are accelerated to energies

$$E_j \approx Ze\beta_j BR_j \text{ erg}, \tag{5.4}$$

where Ze is the particle charge, β_j is the particle velocity over the speed of light, B is the magnetic field in the jet region where the particle is accelerated, and R_j is the jet transverse size. The jet cross section is $S = 3 \cdot 10^{31} \text{ cm}^2$, and the jet Lorentz factor equals 10, according to the active galactic nucleus theory by Vilkoviskij and Karpova (1996). With these parameters the maximum particle energy is

$$E_j \approx 1.9 \cdot 10^{18} \, ZB \text{ eV}. \tag{5.5}$$

Author assumes that the acceleration time T_a is approximately equal to the time T_s during which the particle energy is halved:

$$T_a \approx T_s, \tag{5.6}$$

and that

$$E = E_{\text{max}} \approx 1/2 \, E_j. \tag{5.7}$$

The acceleration time is:

$$T_a \approx l/(\beta_s^2 c), \tag{5.8}$$

where l is the diffusion mean free path, β_s is the shock velocity over the speed of light. In the vicinity of the shock, the value of l is: $l \approx r_L$, where r_L is the particle Larmor radius (see Section 3.3).

The time T_s is:

$$T_s \approx 3.2 \cdot 10^{18} / B_\perp^2 (A/Z)^3 Z^{-1} (Mc^2/E), \tag{5.9}$$

where $B_\perp$ is the field component perpendicular to the particle velocity, A is the particle atomic number, M is the particle mass, $M = Am_p$, and m_p is the proton mass. For most nuclei: $A/Z \approx 2$.

Formulas above are taken from the book by Ginzburg and Syrovatskii (1964) and from the papers by Cesarsky (1992), and Norman et al. (1995).

5.2. THE PARTICLE MAXIMUM ENERGY AND COSMIC RAY CHEMICAL COMPOSITION

Using formulas above author received that in the jet field $B \sim (5\text{-}40)$ G, particles with charges $Z \geq 10$ gain energy of $E \geq 2 \cdot 10^{20}$ eV. Lighter nuclei are accelerated only to $E \leq 10^{20}$ eV, specifically:

- protons gain maximum energy $E_{\max\,p} \approx 3.7 \cdot 10^{19}$ eV in the field $B \approx 19.6$ G;
- He nuclei gain maximum energy $E_{\max\,He} \approx 1.5 \cdot 10^{20}$ eV in the field $B \approx 39.5$ G;
- Fe nuclei gain maximum energy $E_{\max\,Fe} \approx 8 \cdot 10^{20}$ eV in the field $B \approx 16$ G.

In stronger field $B \sim 100$ G particles with $Z > 2$ can gain energy up to $E \geq 10^{20}$ eV. For even stronger field $B \sim 1000$ G, only heavy nuclei with $Z \geq 23$ can be accelerated to $E \geq 10^{20}$ eV.

These estimates are valid for jets with cross sections in the range of $\sim 5 \cdot 10^{29} - 10^{33}$ cm^2.

***

Particles are accelerated at lengths of $(0.1\text{-}20)R_s$ i.e. at lengths $\sim (0.01\text{-}1)$ pc from jet bottom, where $R_s \approx 3 \cdot 10^{14}$ cm is the Schwarzschild radius of the black hole with mass $\sim 10^9 M_\odot$. (The Schwarzschild radius is: $R_s = 2GM^2/c^2$, where G is the gravitational constant, M is the mass of the gravitating object, and c is the speed of light in vacuum.)

At the minimal length $(0.1R_g)$ particles with $Z=23$ are accelerated in the field $B \approx 1000$ G; at the maximal length $(20R_g)$ He nuclei are accelerated in the field $B \approx 39.5$ G. That brings us to the conclusion that for this model the magnetic field near the bottom of the jet is $B \sim 10\text{-}10^3$ G.

***

According to recent measurements by Artukh et al. (2008; 2009) at lengths $\lesssim 1$ pc from the jet bottom the field varies in the range $10^2\text{-}10^4$ G. These experiments confirm the field values obtained from the model.

5.3. ESCAPE OF PARTICLES FROM THE SOURCES

Can accelerated particles escape from sources without significant energy losses? Yes, though there are two processes in which particles can loose energy:

I   Via interactions with infrared (IR) photons,
II   Due to synchrotron radiation and to *curvature radiation* in magnetic fields.

Curvature radiation is emitted by a charged particle moving along curved magnetic lines. Curvature and synchrotron radiations are very similar. The difference is that for synchrotron radiation radius of curvature of the particle trajectory increases for more energetic particles while in curvature radiation the radius is determined only by the magnetic field geometry.

5.3.1. Particle Energy Losses in Interactions with Photons

Regions where the photon density is high are characterized by intense particle-IR photon interactions. These regions include: the accretion disc, the disc funnel, the dust-gas torus surrounding the disc, and the jet. Based on results by Norman et al. (1995) and Kardashev (1995) photolosses in galaxies with luminosity $L<10^{46}$ erg/s and accretion disc optical depth $\tau \leq 1$ are small. In the disc funnel, losses are small when emission has a hard spectrum (Bednarek and Kirk, 1995), which is the case of active galactic nuclei. Particles do not interact with dust in the thick torus if the galactic plane is seen at comparatively small angles.

Significant photolosses result in collimated gamma radiation beams and in roentgen emission of sources. This was discussed by Stecker et al. (1991) and Falcke et al. (1995). On the other hand, sources having negligible photolosses emit small fluxes in roentgen and gamma bands. These are active galactic nuclei that author have identified as cosmic ray sources.

5.3.2. Particle Losses in Magnetic Fields

Since the field is oriented mainly along the jet and the gas flow, i.e. along particle motion energy losses due to synchrotron radiation are negligible.

However lines of forces in dipole magnetic field (magnetic field in sources) curve, and so particles deviate from them and that results in *curvature radiation*.

Author estimates energy losses in curvature radiation, assuming the following model of magnetic fields in sources. First, author assumes that sources are spiral galaxies with typical thickness of ~5 kpc (as our own Galaxy).

Second, author assumes that ultra relativistic particles escape from the galaxy having reached regions where the field is weak, specifically: the particle Larmor radius r_B does not exceed 5 kpc: $r_B \geq 5$ kpc. This condition is fulfilled for particles with maximum energies in regions where $B \leq 10^{-5}$ G. (This is easy to calculate using expression for r_B).

Next, author assumes the field to decrease with distance as $B \sim R^{-3}$ (Vil'koviskii and Karpova, 1996), and the field at $R \sim 1$ pc to be $B \sim 1$ G (Rees, 1987).

Finally, author assumes that particles loose less than one half of the energy in curvature radiation.

With these assumptions author finds the fraction of accelerated particles which escape from the source without significant losses due to curvature radiation. It appears that roughly one in ~300 accelerated particles gets away loosing less than one half of the energy.

***

The curvature radiation losses for a particle with charge Z are given in (Zheleznyakov, 1997):

$$-dE/dt = 2/3(Ze)^2 c(E/Mc^2)^4 (\rho_c)^{-2},$$ (5.10)

where ρ_c is the radius of curvature of the field line. Therefore, particle losses half of energy over time

$$T_{curve} = 7/2(Mc^2)^8 E^{-3}(Ze)^{-2}(\rho_c)^2 c^{-1}.$$ (5.11)

Particle with energy E_{max} traveling distance R_{line} along the field line reaches the end of this distance over time

$$t \approx R_{line}/c \approx 4.6 \cdot 10^9 \text{ s.}$$ (5.12)

Author assumes that curvature radiation losses are small when the particle loses no more than half of its energy E_{max} as it moves along the field line. In this case

$$T_{curve} > t.$$ (5.13)

In dipole magnetic field curvature radius of lines of force in the fixed point equals to $\rho_c = 4R^2/3a$, where R and a are distances between the point and the dipole centre and the dipole axis respectively.

Particles travel to regions where $B \leq 10^{-5}$ G along the field lines, so: $R_{line} = R$ and the maximum particle deviation from the jet axis is a.

Using these assumptions author calculates values of R_{line} and a:

$R_{line} \approx 46$ pc, and

- proton deviation $a_p \approx 0.01$ pc,
- He nucleus deviation $a_{He} \approx 0.03$ pc, and
- Fe nucleus deviation $a_{Fe} \approx 0.04$ pc.

Particles with deviations from the jet axis not exceeding these values leave the source without significant losses due to curvature radiation.

How big is the fraction of these particles?

Angular deviation from the jet axis for these particles equals to

$$\theta \leq a/R_{line} = 6.5 \cdot 10^{-4}, \tag{5.14}$$

and in the shock system particles are scattered isotropically. Based on that author finds the fraction of the particles with small energy losses. In the shock system, the angle θ^* between the velocity vector and jet axis is related to the angle θ (Landau and Lifshitz, 1990):

$$\tan\theta = (1-\beta^2)^{1/2}(\beta + \cos\theta^*)^{-1}\sin\theta^* = 0.14\,\sin\theta^*(0.99 + \cos\theta^*), \tag{5.15}$$

with $\beta \approx 0.99$. For angles $\theta < 0.02$, $\sin\theta^* \approx \theta^*$, $\cos\theta^* \approx 1$ and $\theta \approx 0.01$. The fraction of particles deviating from the axis by angles θ is: $\eta = \theta/\pi \approx 3 \cdot 10^{-3}$, i.e. roughly one in 300 accelerated particles leaves the source without losses in curvature radiation.

5.3.3. Energy Losses which Were Not Accounted for

Jet can also excite *bow shocks* in the hot gas flow. When the gas density of the flow is lower that that of the jet, the bow shock propagates slower than jet. In this case particles do not interact with the bow shock and travel away without energy losses. In the opposite case when the density of jet is higher than that of the gas, particles loose energy in the bow shock. Author does not account for losses of this kind. Jet propagation and its structure are described by S. Chakrabarti (1988).

5.4. CONCLUSION

Proposed model describes particle acceleration in low-luminosity Seyfert nuclei. This model is based on the following assumptions.

I   The acceleration takes place in relativistic jets, namely at shock fronts with a regular magnetic field.

II   Shocks are exited on the jet surface when jet propagates through the matter.

III Jets contain material from the accretion disc, so there are protons and atomic nuclei (or their fragments).

IV   Since magnetic field in jets is not well known it is treated as a parameter and fields are considered in the range $B \sim 5\text{-}1000$ G.

V   Values for cross section and Lorentz factor for the jet are taken from (Vil'koviskii and Karpova, 1996).

According to the discussed model, particles are not only accelerated to ultra high energies in the sources, but they also can travel away from the galaxies without significant energy losses.

This model predicts that both the cosmic ray maximum energy and element composition depend on the jet magnetic field. The highest energy $E \approx 8 \cdot 10^{20}$ eV is obtained for Fe nuclei with the field in the jet $B \approx 16$ G. Protons are accelerated only to energies $E < 4 \cdot 10^{19}$ eV for any value of B.

In general in the field $B \sim (5\text{-}40)$ G, nuclei with $Z \geq 10$ are accelerated to energies $E \geq 2 \cdot 10^{20}$ eV, while lighter nuclei to energies $E \leq 10^{20}$ eV. In the field $B \sim 1000$ G, only heavy particles with $Z \geq 23$ can obtain energies $E \geq 10^{20}$ eV.

Values of the jet magnetic field predicted by the model appear to be in agreement with recent observational data. In observed active galactic nuclei field ranges from 10^2 to 10^4 G at distances less than 1 pc from bottom of the jet (Artyukh et al. 2008; 2009).

Also the maximum particle energy in this model $E \sim 10^{21}$ eV coincide with the maximum energies detected in cosmic rays by Bird et al., (1995).

Therefore the suggested model predicts that cosmic rays can be accelerated to ultra high energies in low-luminosity Seyfert nuclei. Moreover, author concludes that cosmic ray chemical composition can be used to estimate magnetic fields of jets. Finally author concludes that detected ultra high energy protons are either atomic nuclei fragments or they have been accelerated in other objects.

6. MODEL FOR COSMIC RAY ACCELERATION IN BL LAC OBJECTS

The model for particle acceleration in objects like BL Lac's was proposed by Kardashev (1995) and further investigated by Shatskii and Kardashev (2002). Author uses it to analyze cosmic ray data. Below is a brief description of this model.

Following (Kardashev, 1995) the particle acceleration occurs in the electric field induced near a supermassive black hole with the mass $\sim 10^9 M_\odot$ in the periods of the black hole low activity. During these periods the accretion is diminished. The model is based on two main assumptions. The first is that the magnetic field of the black hole can be as high as $B \sim 2 \cdot 10^{10}$ G. Second, it is assumed that when the black hole is in the quiescent states of its activity there is a region with a very low plasma density, in which a very strong induced electric field is not compensated by the volume charge of plasma.

The value of the magnetic field assumed in this model is in contrast with the value $B \sim 10^4$ G derived by Field and Rogers (1993), and the limit of field strength derived by Krolik (1999): $B^2 \sim 8\pi\rho$, where ρ is the density of matter in the accretion disk. However Zakharov et al. (2003) interpreting recent observations demonstrated that magnetic fields as strong as $B \sim 2 \cdot 10^{10}$ G exist in active galactic nuclei.

Regions in which super strong electric fields can exist are located near the magnetic poles and the rotational axis of the black hole. In these fields particles with charge Z can be accelerated to energies $E \sim 10^{27} Z$ eV. The accelerated particles lose energy in interactions with photons in the disk and in magnetic fields via curvature radiation. In quiescent phases, losses in photon reactions are negligible. On the other hand losses in magnetic fields are significant: the particle energy can be decreased to $10^{21}Z$ eV due to curvature radiation, and therefore the maximum proton energy is $E\sim10^{21}$ eV.

This model allows particles to accelerate in the presence of extremely strong electric fields. The maximum particle energies $E\sim10^{21}$ eV is in agreement with the maximum energy achievable in the Seyfert nucleus for the model discussed in Section 5.

7. PARAMETERS OF PARTICLE ACCELERATION IN ACTIVE GALACTIC NUCLEI OBTAINED FROM COSMIC RAY DATA

In this Section author demonstrates that using cosmic ray data it is possible to derive some characteristics of particle acceleration in active galactic nuclei, specifically: cosmic ray spectra in the sources and the source luminosity in cosmic rays.

7.1. INITIAL COSMIC RAY SPECTRA IN ACTIVE GALACTIC NUCLEI DERIVED FROM MEASURED SPECTRA

There are two independent mechanisms of particle acceleration that are realized in cosmic ray sources: first, acceleration in the electric field and second, acceleration on shock fronts in presence of a regular magnetic field. The former is the mechanism suggested for BL Lac's by Kardashev (1995). The latter is used in the model of particle acceleration in Seyfert nuclei discussed above.

Using the first mechanism, author assumes the simplest case: the particle initial energy spectrum is monoenergetic, with the energy $E \sim 10^{27} Z$ eV or $E \sim 10^{21} Z$ eV.

In the second case, when particles are accelerated on shock fronts, the particle initial energy spectrum is exponential. As it was shown above this acceleration mechanism can be realized in low-luminosity Seyfert nuclei.

So AUTHOR assumes that the particle energy spectrum at the source is either monoenergetic or exponential.

Traveling particles loose energy via GZK-effect, and also adiabatically due to the Universe inflation. As a result energy spectrum of cosmic rays near Earth differs from the initial spectrum in the sources. It would be interesting to calculate cosmic ray spectrum after particle propagation in the extragalactic space and to compare calculated spectrum with a measured one.

7.1.1. Calculation of the Cosmic Ray Energy Spectra Near the Earth

Author calculates cosmic ray energy spectra near the Earth using following assumptions. First, author uses active galactic nucleus distributions in red shifts z according to catalogue data. These distributions for Seyfert nuclei and BL Lac's are shown in figures 4 and 5, respectively. Author adopts that the distance r to an object with red shift z equals

$$r = 2/3cH^{-1}[1-(1+z)^{-3/2}] \text{ Mpc,} \qquad (7.1)$$

and that according to the Einstein–de Sitter model of the Universe time and red shift are related as

$$t = 2/3H^{-1}(1+z)^{-3/2} . \qquad (7.2)$$

In the Einstein–de Sitter model parameter Ω is used to characterize the inflation of the Universe. Formulas (7.1) and (7.2) were derived in the case $\Omega=1$ when inflation stops. May be this is not correct as recent astronomical observations show. However $\Omega=1$ is suitable for the problem of cosmic ray propagation.

Second, there are two types of the cosmic ray initial spectra for both Seyfert nuclei and BL Lac's. They are either monoenergetic or exponential.

Next, author assumes that cosmic ray particles are protons. This is true for particles generated by BL Lac's, because these sources are so distant that any nucleus disintegrates relatively close to the source due to interactions with IR photons. However for cosmic rays emitted by Seyfert nuclei this is a simplification. Below author considers two types of proton interactions – with relic and IR protons.

Author takes into account two factors resulting from the inflation of the Universe. First, traveling protons lose energy adiabatically. Second, at the epoch with a red shift z the relic photon density and energy were higher than at $z=0$. Following Berezinsky et al. (1990) author assumes them to be $(1+z)^3$, $(1+z)$ times greater than at $z=0$ respectively.

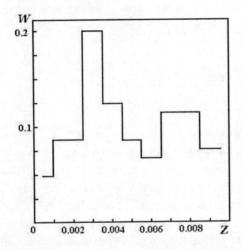

Figure 4. Red shift distribution of nearby ($z \leq 0.0092$) Seyfert galactic nuclei normalized to their total number.

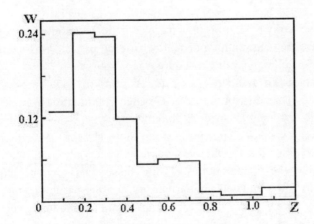

Figure 5. Red shift distribution of BL Lac objects normalized to their total number.

***

Relic photon spectrum is described by Planck's law. In calculation the spectrum was divided into two parts. Most photons assumed to have following parameters: temperature - $T=2.7$ K, photon energy - $\varepsilon \approx 6 \cdot 10^{-4}$ eV, and photon density - $n_0 \approx 400$ cm^{-3} (which are mean parameters in Planck's law). For high-energy photons, energy is $\varepsilon_t \approx 1 \cdot 10^{-3}$ eV and the density is $n_t \approx 42$ cm^{-3}.

The energy range of the infrared radiation is $2 \cdot 10^{-3}$–0.8 eV. Following Puget et al. (1976) and Stecker (1998) author uses formula for infrared radiation spectrum:

$$n(\varepsilon) = 7 \cdot 10^{-5} \varepsilon^{-2.5} \text{ cm}^{-3} \text{ eV}^{-1}. \tag{7.3}$$

The mean energy of the infrared photons is $\langle \varepsilon_{IR} \rangle \approx 5.4 \cdot 10^{-3}$ eV, and their mean density - $\langle n_{IR} \rangle \approx 2.28$ cm^{-3}.

The photopion reactions are threshold ones with threshold energy $\varepsilon_{th}^* \approx 145$ MeV, where ε^* is photon energy in the proton system. Reaction cross section σ and the coefficient of inelasticity K depend on the energy ε^*. The data for $\sigma(\varepsilon^*)$ and $K(\varepsilon^*)$ were taken from (Stecker, 1968) and (Particle Data Group, 2002).

Adiabatic losses of a proton with an initial energy E which travels from a point with a red shift z to a point with $z = 0$ are:

$$-dE/dt = H(1 + z)^{3/2}E. \tag{7.4}$$

Finally, cosmic ray deviations in extragalactic fields assumed to be negligible.

With these assumptions propagation of each particle was simulated using Monte-Carlo technique. The procedure was as follows.

Simulation for red shift z of the cosmic ray source was performed considering source distribution. Particle energy simulation was performed for the case of exponential spectra. Photon energy was simulated considering photon energy spectra. After performing these simulations author calculated value of cross section for the particle-photon interaction and particle energy losses during these interactions. Using these data author obtained mean particle path and simulated photon density and the particle path. In this way GZK-losses were calculated for each point where interaction occurred. Particle losses caused by the Universe inflation were calculated using formula: $(\Delta E)_{ad} = E(z_i - z_{i+1})/(1 + z_i)$, where z_i, z_{i+1} are red shifts of points of two subsequent interactions (for the first interaction z_i equals the source red shift). These losses were added up GZK-losses.

7.1.2. Calculated Spectra and Comparison with Experiment

Obtained cosmic ray energy spectra are shown in figures 6 and 7. Author normalized them to the spectra measured at the energy $E \approx 5 \cdot 10^{19}$ eV.

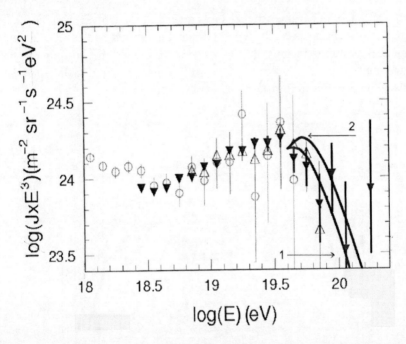

Figure 6. The cosmic ray differential energy spectrum got at Pierre Auger observatory (2007b). Black and white triangles denote spectra got using vertical and inclined showers, respectively. Circles denote the spectrum got with hybrid data set. Solid lines show cosmic ray calculated spectra: (1) – sources are BL Lac's with the power-law initial spectrum, $\chi=3.0$; (2) sources are BL Lac's with the power-law initial spectrum, $\chi=2.6$.

Comparing calculated spectra with experimentally obtained we should keep in mind that spectra obtained at various arrays commonly do not coincide with each other and measurement errors are big. Nevertheless author performed this comparison and came to well-defined conclusions. They are formulated as follows.

First, in sources maximum energy of accelerated particles does not exceed $E \approx 10^{21}$ eV. Aharonian et al. (2002) and Medvedev (2003) obtained this limit theoretically.

Second, the Seyfert model with the monoenergetic initial cosmic ray spectra is evidently in contradiction with data of all arrays.

As for initial monoenergetic cosmic ray spectra in BL Lac's the model fits spectra experimentally observed at Fly's Eye, Haverah Park, and Yakutsk arrays. However model with initial power spectrum fits Pierre Auger and HiRes spectra.

Because of experimental errors it is difficult to reliably derive the value of exponent. Nevertheless author concludes that in Seyfert nuclei the value of the exponent is approximately 3.0. (This value provides the best fit for Pierre Auger and HiRes data). For BL Lac's, the value of exponent is either 2.6 or 3.0 and both seems to fit data of observations.

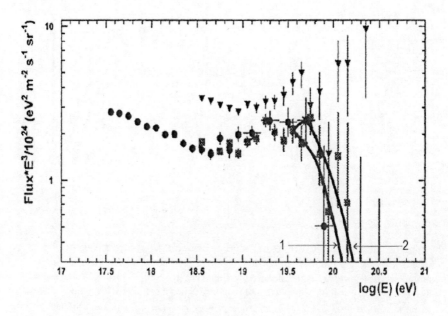

Figure 7. The cosmic ray differential energy spectrum got at the HiRes array (2007). Squares denote HiRes-1 data, circles denote HiRes-2 data, and triangles denote AGASA data. Solid lines show cosmic ray calculated spectra: (1) – sources are BL Lac's with the power-law initial spectrum, $\chi = 3.0$; (2) sources are BL Lac's with the power-law initial spectrum, $\chi = 2.6$.

7.2. ESTIMATE OF SEYFERT NUCLEUS LUMINOSITY IN COSMIC RAYS

Author proceeds with estimate of the source luminosity which is observed in ultra high energy cosmic rays.

Following Berezinsky et al (1990) author assumes that luminosity in cosmic rays equals

$$L_{UHECR} = \int_{E}^{\infty} F_g(E)E dE, \qquad (7.5)$$

where $F_g(E) = KE^{-\gamma}$ is the particle differential spectrum in the source. Author assumes that the initial spectrum is weakly distorted through interactions in extragalactic space, sources being located at distance $L \lesssim 50$ Mpc around Earth. Because of weak distortion the spectrum $I(E)$ observed at $E > 5 \cdot 10^{19}$ eV is also exponential: $I(E) \approx KE^{-\gamma}$, $\gamma \geq 3.1$. Next, in the Universe total cosmic ray intensity is

$$I(E) = (c/4\pi)F_g(E)n_{Sy}T_{CR}, \qquad (7.6)$$

where $n_{Sy} = 2 \cdot 10^{-77}$ cm^{-3} is Seyfert nucleus volume density and T_{CR} is the cosmic ray age. Assuming sources at distances $L \approx 50$ Mpc, the cosmic ray age is

$$T_{CR} = L/c \approx 1.7 \cdot 10^8 \text{ years.} \qquad (7.7)$$

Different arrays give the cosmic ray intensity (Watson, 1995):

$$I(E) \approx 10^{-39} - 10^{-40} \text{ (cm}^2 \cdot \text{s} \cdot \text{sr} \cdot \text{eV)}^{-1}. \qquad (7.8)$$

Using formulas (7.5) – (7.8) author calculates Seyfert nucleus luminosity in cosmic rays: $L_S \approx 10^{41} - 10^{42}$ erg/s if $\gamma \approx 3.1$ and $L_S \approx 10^{39} - 10^{40}$ erg/s if $\gamma \approx 3.0$.

The real cosmic-ray luminosity of the source can be greater than that observed by $1/\eta$ times (η is defined in Section 5.3.2).

Assuming that the mass of a black hole is $M{\approx}10^9 M_\odot$, and all its energy is used for cosmic ray acceleration author estimates that the Seyfert nucleus energy will be depleted in approximately $10^{13} - 10^{14}$ years. This period is very long compared to the age of the Universe $T_U{\approx}1.3{\cdot}10^{10}$ years. That brings us to the conclusion that low-luminosity Seyfert nuclei store enough energy for particle acceleration and they are capable to accelerate cosmic particles for a long time.

7.3. ESTIMATE OF BL LAC LUMINOSITY IN COSMIC RAYS

Author estimates the BL Lac luminosity in cosmic rays in the following way.

The luminosity L_{BL} is equal to

$$L_{BL}=U_{CR}/(N{\cdot}T), \tag{7.9}$$

where U_{CR} is the total energy emitted in cosmic rays, N_{BL} is the number of BL Lac's, and T is the cosmic ray age.

The value of U_{CR} can be got from the energy balance relation:

$$U_{CR} =(U_{CR})_{measured} + (U_{CR})_{lost}, \tag{7.10}$$

where $(U_{CR})_{measured}$ is the energy of cosmic rays registered and $(U_{CR})_{lost}$ is the cosmic ray energy which was lost during particle propagation.

Author assumes the simple case that the cosmic ray initial energy is $E_0=10^{21}$ eV and that the bulk of cosmic rays registered has the energy $E=5{\cdot}10^{19}$ eV. Under these assumptions author gets:

$$(U_{CR})_{measured}/U_{CR}{\approx}E/E_0{\approx}0.05 \text{ and } U_{CR} {\approx}20(U_{CR})_{measured}. \tag{7.11}$$

Now author gets the value of $(U_{CR})_{measured}$. Author defines $(U_{CR})_{measured}$ as

$$(U_{CR})_{measured}= \int\limits_{E} I(E)EdE \cdot 4\pi/c{\cdot}V, \tag{7.12}$$

where $I(E)$ is the cosmic ray intensity, V is the volume filled by cosmic rays. Author estimates the integral using the ultra high energy cosmic ray intensity $I(E)$ measured. The integral equals approximately 4 eV/(cm^2·s·sr).

The majority of BL Lac's has red shifts $z{\leq}0.35$, hence they are located at distances $r{\leq}1000$ Mpc. Assuming that cosmic rays fill the sphere with the radius $r{\approx}1000$ Mpc author gets the value V and then $(U_{CR})_{measured}$.

Cosmic ray sources filling the above sphere, cosmic rays reach Earth during $T{\leq}2{\cdot}10^{17}$ s. Assuming that cosmic rays reach us for $T{\approx}2{\cdot}10^{17}$ s author gets

$$U_{CR}/T{\approx}2{\cdot}10^{44} \text{erg/s.} \tag{7.13}$$

The number of BL Lac's at $z{\leq}0.35$ is $N_{BL}{\approx}100$, hence $L_{BL}{\approx}2{\cdot}10^{42}$ erg/s. The number of BL Lac's can be greater, and then the source luminosity is lower than the value L_{BL} obtained.

7.4. CONCLUSION

Author considers two main processes of charged particle acceleration: by electric fields and on shock fronts with regular magnetic fields. Next, author supposes that initial cosmic ray spectra in sources are of two types: monoenergetic, and exponential. It is demonstrated that based on measured cosmic ray spectrum we can distinguish between these two types of spectra.

At present errors in the measured spectra are big. In addition not all data of arrays coincide. Pierre Auger and HiRes data suggest an exponential cosmic ray spectrum in sources.

Using the cosmic ray data following parameters of cosmic ray sources are got: the value of maximal particle energy in sources which equals $E{\approx}10^{21}$ eV and the value of source luminosity in cosmic rays.

8. COSMIC RAY CLUSTERS
AND VARIABILITY
OF SOURCES

8.1. WHAT ARE CLUSTERS
IN COSMIC RAYS?

AGASA data contain 58 ultra high energy particles including 5 doublets and 1 triplet in which particle arrival coordinates coincide within 1-error box. These data were collected over the period of ten years. The list of clusters is published by Hayashida et al. (2000) where they demonstrate that arrival coincidences are not accidental. In addition, one ultra high energy particle registered by Yakutsk array has arrival coordinates in the error-box of one of the clusters. In clusters particle registration frequency is approximately (1-1.5) year^{-1}, however in one of doublets particles were registered during 10 years.

What is cluster origin? This problem is considered in this Section.

8.2. ESTIMATES OF COSMIC RAY
REGISTRATION FREQUENCY

First consider the case of cosmic rays generated by a single source. How often particles with coinciding coordinates can be generated by a source?

Author considers arrays first with area $S \approx 10$ km^2 (approximately the area of Yakutsk, Haverah Park, Akeno, and Volcano Ranch arrays) and second with areas $S \approx 100$ and $S \approx 3000$ km^2 (areas of the AGASA and Pierre Auger arrays). Author estimates the period of time over which a cluster can be registered by these arrays. (For details see Uryson (2005).)

8.2.1. Clusters from a Single Seyfert Nucleus

In Section 7, author supposed that host galaxies with Seyfert nuclei emitting cosmic rays oriented so that the galactic plane is seen at comparatively small angles.

However Seyfert galaxies identified as possible cosmic ray sources are oriented differently: the mean angle between the light of sight and the normal to the galactic plane is $i \approx 52^0$. Among Seyfert nuclei with $z < 0.01$, the fraction of sources that satisfy the model is only 0.15, and author believe that this low number cannot explain the existence of clusters. Therefore author assumes that cosmic rays are emitted by Seyfert nuclei which belong to host galaxies of any orientation.

One of similar models is suggested by Haswell et al. (1992), who supposed that particles are emitted by the accretion disk.

With this in mind author considers a single Seyfert nucleus with the cosmic-ray luminosity $L_S \approx 10^{40}$ erg/s. Author assumes that this active nucleus is located at $R \approx 16$ Mpc from the Earth. Author uses the value $R \approx 16$ Mpc because at this distance space distribution of Seyfert nuclei with $z < 0.01$ has the maximum (assuming Hubble constant $H = 75$ km/(Mpc s)).

Using these parameters, author calculates that cosmic ray flux near Earth is equal to $N \approx 4.3 \cdot 10^{-21}$ particle/(cm^2 s). If this is the case then arrays with areas $S = 10$, 100, and 3000 km^2 should detect 0.012, 0.12, and about 4 particle per year, respectively. Therefore an array with area $S = 10$ km^2 will register a doublet in more than 100 years of observations. An array with $S \approx 100$ km^2 (AGASA array) can register a doublet over a period of $T > 10$ years. An array with $S \sim 3000$ km^2 (Pierre Auger observatory) can detect a cluster in just one year.

However there were no clusters detected at Pierre Auger observatory. So what is wrong with estimates? Above author assumed that the cosmic-ray luminosity is $L_S \approx 10^{40}$ erg/s. But this is an estimate that depends on parameters of the model and can be within the range from 10^{39} to 10^{42} erg/s. (This was discussed in Section 7.3.) If a smaller value is used for the cosmic-ray luminosity than clusters can be registered over longer period of time. For example if the source luminosity in cosmic rays is $L_S \approx 10^{39}$ erg/s a cluster can be registered over 10 years of observations.

8.2.1. Clusters from a Single BL Lac

BL Lac's have powerful jets directed towards Earth. Author accepts that particles are emitted in jets that oriented towards us. Considering BL Lac's, author assumes that the initial cosmic ray spectrum is monoenergetic with the energy of 10^{21} eV. Also it is assumed that the source has the red shift $z \approx 0.2$ (this is the red shift corresponding to the maximum of the BL Lac distribution) i.e. the source is at $R \approx 600$ Mpc away.

***

Near Earth, the fraction of particles with energies $E \geq 4 \cdot 10^{19}$ depends on the distance from the source due to the GZK-effect. For distances $R \approx 600$ Mpc this fraction is 0.7. For illustration, fractions of 0.3, 0.5, and 1 correspond to distances $R \approx 800$, 700 Mpc and $R < 600$ Mpc respectively. Contribution of sources to the cosmic ray flux at $R > 800$ Mpc is negligible. It is an order lower than the contribution of sources at $R < 600$ Mpc due to the low density of particles emitted by distant sources irrespectively of the GZK-effect.

***

Following model suggested by Kardashev (1995) author assumes that the beam aperture angle is $\alpha \approx 1.4 \cdot 10^{-6}$.

For power L_{BL} that is spent on cosmic ray emission author considers two values: $L_{BL} \approx 6 \cdot 10^{45}$ erg/s and $L_{BL} \sim 10^{43}$ erg/s. The former value is derived from (Kardashev, 1995) and the latter one is estimate of BL Lac luminosity in cosmic rays derived in Section 7.4.

Author adopts two assumptions: first, power L_{BL} is emitted inside the beam and second, it is emitted isotropically.

Using first assumption the model predicts that arrays with area $S \approx 10$ km^2 detect $\sim 10^8$ and 10^6 particle/year from a single source, respectively. This is in conflict with experimental data and author excludes these cases from the following discussion.

Next, author supposes that $L_{BL} \approx 6 \cdot 10^{45}$ erg/s is the power of isotropic emission. In this case according to the model an array with area $S \approx 100$ km^2 registers 1.8 particle/year, so in several years it detects a cluster with at least 3 particles. An array with $S \sim 10$ km^2 can detect a doublet over a period of observation of $T > 10$ years.

Finally, assuming the power of isotropic emission is $L_{BL} \sim 10^{43}$ erg/s as obtained in Section 7.4, only arrays with area $S \sim 3000$ km^2 can detect clusters, over a period of 10 years and shorter.

8.2. AGASA Clusters
and Active Galactic Nucleus Model

Author assumes that particles forming clusters come from regions with higher local space densities of active galactic nuclei. Based on that author estimates how many active galactic nuclei are located inside the 3-error boxes of coordinates of particles forming the cluster.

It appears that the model that describes "nearby" Seyfert nuclei with $z < 0.01$ as a source of cosmic ray particles explains the origin of two clusters (out of five). Origin of four clusters is described by the model that adopts BL Lac's as sources of particles. In the model for BL Lac's clusters also can be emitted by single sources.

One doublet does not fit in any model. This doublet was formed by particles arrived from the galactic latitude $b \approx -10^0$, which is in an "avoidance zone". (Dubovsky et al. (2000) suggested another explanation of cluster origin.)

8.3. Factors which Were Not Considered

Actual observation periods can be longer than the above estimates due to a number of reasons.

First, cosmic ray fluxes were derived assuming that an array detects particles from a source during its total operating time. In reality source location in the sky depends on the season of the year and on the time of day. Because of this an array detects the source emission during shorter periods. For example, if an array observes a sky region for ~1/2 day during 1/2 year the observation period and consequently the registered flux are actually four times lower.

Second, usually showers with axes deviating from the vertical within angles $\Theta < (30\text{-}45^0)$ are selected. Due to this an array detects only about a half of particles emitted by the source.

Therefore the actual registration frequency can be approximately (2-10) times higher than the estimated one.

These two factors result from array operation. In addition, there are also physical reasons by which cosmic ray registration frequency can differ from estimates above.

Certain intrinsic features of the source can also influence formation of a cluster. First, active galactic nuclei activity is not permanent, but is quasi periodic. As revealed by Pyatunina et al. (2007) active galactic nuclei are active during some years, after which they seem to be inactive for about 25 years. From this it is clear that particles are accelerated in the source during a period of some years, after which the source is off. The next cycle of activity takes place in about 20-30 years. Second, fraction of protons and atomic nuclei in plasma blobs in jet varies. Author supposes that this fraction ranges from ~0.01 to ~0.1. (This value is not in conflict with results by Zheleznyakov and Koryagin (2002).)

Source switching appears to be one of the reasons why so far there have been detected no clusters with particle number greater than three at AGASA.

8.4. CONCLUSION

Our model describes the origin of particle clusters detected at AGASA. According to this model cosmic ray clusters arrive from celestial regions with higher volume densities of active galactic nuclei. In equatorial coordinates sizes of these sky regions are ($\Delta\alpha < 9^0$, $\Delta\delta < 9^0$). Also clusters can arrive from single active galactic nuclei. In this case cosmic ray sources are BL Lac's. According to the model, no clusters can be detected at arrays with areas $S \sim 10$ km^2.

Variability in source activity (and in composition of jet plasma blobs as well) result in cosmic ray switching on and shutdown. Breaks in activity can be as long as approximately 25 years. So in order to register a cluster both longer exposure times and larger area arrays are required.

9. INVESTIGATION OF EXTRAGALACTIC RADIO BACKGROUND USING COSMIC RAY DATA

9.1. INTRODUCTION

In the extragalactic space ultra high energy cosmic rays interact with microwave background photons. This interaction is known as GZK-effect. It was briefly discussed in Section 2.3.2. GZK-effect appears in the shape of the cosmic ray energy spectrum on Earth and in electromagnetic cascades in extragalactic space. Form of the spectrum was studied by Hillas (1984), Hill and Schramm (1985), and Berezinsky and Grigor'eva (1988). Hayakawa (1966) and Prilutskiy and Rozental (1970) discovered and examined electromagnetic cascades producing electrons and gamma rays in the extragalactic space. Akerlof et al. (2003) attempted to discover cascade gamma rays at energies E>350 GeV, but they did not succeed.

However most of the gamma rays are originated in other processes. First, they are produced in the Galaxy via cosmic ray interactions with interstellar gas at particles energies $E<10^{15}$ eV. Second, at these energies pulsars also are sources of gamma rays. In addition, active galactic nuclei are also identified as possible gamma ray sources. A lot of gamma ray sources seems to be unresolved. These sources contribute to gamma-ray background. How can we select quanta generated in cascades from quanta of other origin? For the first process, the intensity of gamma rays emission reaches its maximum in the galactic plane. Discrete sources being gamma ray emitters, there are peaks in the gamma ray intensity toward emitters. Contribution of unresolved sources is estimated theoretically. Using this data it is possible to select cascade gamma quanta from those of other origin.

In this Section author studies initial energy CR spectra in sources and the extragalactic background radio spectrum using gamma rays originated in extragalactic cascades.

How does the intensity of cascade gamma quanta depend on initial cosmic ray spectra in AGN and on the extragalactic background radio spectrum? To answer this question author calculates the cascade gamma ray intensity at energy $E \approx 10^{14}$ eV. This range is chosen because according to (Gould and Schreder, 1967) the Universe is practically opaque for quanta at $E \approx 10^{14}$ eV and the absorption is minimal. Author considers different models of CR sources and different estimates of extragalactic radio background intensity.

Based on the results author suggests to study initial energy CR spectra in sources and the extragalactic radio background by detecting cascade gamma rays with energy $E \approx 10^{14}$ eV.

9.1. MAIN FEATURES OF EXTRAGALACTIC CASCADES

In this Section author assumes that cosmic rays consist of protons.

Extragalactic cascades are initiated and develop through following processes. First, protons at energies $E > 4 \cdot 10^{19}$ eV interact with relic photons:

$$p + \gamma_{rel} \rightarrow p + \pi^0, \tag{9.1}$$

$$p + \gamma_{rel} \rightarrow n + \pi^+. \tag{9.2}$$

Cross sections of these reactions depend on energies of interacting particles with peak at ≈ 0.44 mb for photon energy $\varepsilon^* \approx 340$ GeV in the proton rest system (Particle Data Group, 2002).

Pion decays give rise to electrons and photons

$$\pi^0 \rightarrow 2\gamma, \tag{9.3}$$

$$\pi^+ \rightarrow \mu^+ + \nu, \tag{9.4}$$

$$\mu^+ \rightarrow e^+ + \nu + \nu, \tag{9.5}$$

which in turns originate the cascade via reactions with background photons. These reactions include production of pairs

$$\gamma + \gamma_b \rightarrow e^+ + e^-, \qquad (9.6)$$

and inverse Compton scattering

$$e + \gamma_b \rightarrow e' + \gamma'. \qquad (9.7)$$

Pair production takes place provided that the photon energy is above the threshold energy

$$E_i = (mc^2)^2 / \varepsilon_b, \qquad (9.8)$$

where mc^2 is the electron mass, $mc^2 = 0.511$ MeV, and ε_b is the energy of the background photon.

***

Contributions of two other processes: $\gamma + \gamma_b \rightarrow e^+ + e^- + e^+ + e^-$ and $e + \gamma_b \rightarrow e' + e^+ + e^-$, are negligible. The former dominates at energies $E_\gamma \geq 10^{21}$ eV, whereas the maximal secondary gamma ray energy in this model is $E \approx 1.5 \cdot 10^{20}$ eV. Contribution of the latter process is negligible because energy transfer to pairs of particles is smaller than 10^{-3} (Bhattacharja and Sigl, 2000).

Cross sections of the processes are well known. For pair production it is

$$\sigma_{\gamma\gamma} \approx 3/8 \sigma_T a^2 / \{[2 + 2a^2 - a^4] \ln(a^{-1} + (a^{-2} - 1)^{1/2}) - (1 - a^2)^{1/2}(1 + a^2)\}, \quad (9.9)$$

where $a = mc^2 / E_{ec}$, E_{ec} is the photon energy in the centre-of-mass system

$$E_{ec} = [E_e \varepsilon_b (1 - \cos \psi)]^{1/2}, \qquad (9.10)$$

ψ is the angle between photon impulses in the laboratory system.

The cross section of the inverse Compton scattering equals

$$\sigma_{IC} \approx 3/8 \sigma_T (mc^2)^2 / (E_e \varepsilon_b) \{\ln[2 E_e \varepsilon_b / (mc^2)^2] + 0.5\} \text{ at } E_e > E_t; \qquad (9.11)$$

at $E_e < E_t$ it is equal to Thomson cross section, $\sigma_T \approx 6.65 \cdot 10^{-25}$ cm^2.

***

If energy of one of the primary particles (γ or e) is above the threshold of one of the reactions i.e. E_γ, $E_e > E_t$, one of the pair particles and the scattered photon receive almost all the primary energy (E_γ or E_e respectively). Another particle in the pair and the scattered electron receive energy of about E_t.

So at energies above threshold high energy gamma quantum converts into a high energy particle of a pair, after which this particle converts into the gamma quantum via inverse Compton scattering, and so forth. Therefore in the cascade there is a particle which carries about the same energy as the primary proton. This particle is called "leading particle.

At lower primary particle energies, $E_e < E_t$, there is no pair production, and inverse Compton scattering produces comparatively soft photons with mean energies

$$E_\gamma = 4/3 \varepsilon_b (E_e / mc^2)^2. \tag{9.12}$$

In this case electron looses energy fairly slowly.

9.2. ELECTRON SYNCHROTRON EMISSION IN EXTRAGALACTIC MAGNETIC FIELDS

Since electrons in magnetic fields loose energy via synchrotron radiation this process can stop development of cascades. Below author demonstrates that extragalactic magnetic fields are too weak to violate cascade developing. This can be done using two approaches.

In the first approach, synchrotron losses are small compared to losses in the inverse Compton scattering

$$(dE/dt)_s < (dE/dt)_{IC}. \tag{9.13}$$

Mean energy losses of electrons in these processes are given by the following expressions (Ginzburg and Syrovatskii, 1964):

- electron synchrotron losses

$$-(dE/dt)_s = 6.5 \cdot 10^{-4} B^2 (E/mc^2)^2 \text{ eV/s}, \tag{9.14}$$

- losses via the inverse Compton scattering

$$-(dE/dt)_{IC}=1.65 \cdot 10^{-2}\, w_b (E/mc^2)^2 \text{ eV/s at } E_e < E_t, \qquad (9.15)$$

$$-(dE/dt)_{IC} \approx 6.2 \cdot 10^{-3} (mc^2/<\varepsilon_b>)^2\, w_b \ln(2E<\varepsilon_b>/m^2 c^4) \text{ eV/s at } E_e > E_t. $$

Second approach assumes that electrons scatter hard photons faster than they loose a half of energy in synchrotron emission:

$$\Delta t_{IC} < T_s, \qquad (9.16)$$

where Δt_{IC} is duration of scattering and T_s is duration of process of synchrotron emission in which electron energy reduces in two times. The value of T_s is given by (Ginzburg and Syrovatskii, 1964):

$$T_s = 5 \cdot 10^8\, mc^2/(B_\perp^2 E_e), \qquad (9.17)$$

here $B_\perp$ in G is the field component normal to electron velocity, and the electron energy is in eV.

The value Δt_{IC} equals

$$\Delta t_{IC} = \lambda_{IC}/c, \qquad (9.18)$$

where λ_{IC} is electron mean free path, $\lambda_{IC}=(\sigma_{IC} \cdot n_b)^{-1}$, n_b is the background photon density and c is the speed of light.

Estimates show that both conditions (9.13) and (9.14) are satisfied in the magnetic field $B \approx 10^{-11}$ G. Hence ultra high energy electrons have minor synchrotron losses in the magnetic field $B \approx 10^{-11}$ G. For electrons with $E \approx 10^{14} - 10^{17}$ eV, synchrotron losses are minor in the field $B < 2 \cdot 10^{-8}$ G. Finally, electrons at $E \approx 10^{14}$ eV have negligible synchrotron losses in the magnetic field $B < 5 \cdot 10^{-6}$ G. (The energy value of 10^{14} eV is the lowest energy which I assume.)

These limits coincide with estimates of extragalactic field derived in Sections 3 and 4. Hence, the extragalactic field is too weak to violate cascade developing. In the Galaxy, the field is $B \approx 3 \cdot 10^{-6}$ G, and electrons loose energy in synchrotron emission.

9.3. EXTRAGALACTIC BACKGROUND EMISSION

The role of the extragalactic background emission is of greatest importance for cascade development.

Relic photons have Planck distribution. They were treated in the way described in Section 7.1.

Data about extragalactic radio background are not complete. Clark et al. (1982) experimentally measured radio photon energy and came to the conclusion that it is higher than $\varepsilon_b \approx 2 \cdot 10^{-8}$ eV, with densities $n_b \approx 0.1 \mathrm{cm}^{-3}$. However estimates by Protheroe and Biermann (1997) demonstrate that radio photons exist at much lower energies $\varepsilon_b \approx 4 \cdot 10^{-10}$ eV, with densities $n_b \approx 1 \mathrm{cm}^{-3}$.

Following formula (9.8) for threshold energy, the lower the background photon energy ε_b, the greater is the value of E_t. The most low-energy background radiation in the Universe is the non thermal radio emission. In this case the threshold energy should be $E_t \geq 10^{19}$ eV, specifically, $E_{t1} \approx 1.5 \cdot 10^{19}$ eV, for radio photon energies according to (Clark et al. 1982), and $E_{t2} \approx 6 \cdot 10^{20}$ eV using estimates by (Protheroe and Biermann, 1997). Interacting with relic photons requires threshold energy of several orders of magnitude lower $E_t \approx 10^{14}$ eV.

***

The question to conclude the qualitative analysis is: what energies are more effective for cascade developing?

Electrons and quanta at sufficiently high energies interact with radio photons via processes (9.6) and (9.7). If energies of both produced particles are sufficient the cycle of pair production continues via the same processes. If the energy is high enough the cycle proceeds through interactions with radio photons, and at lower energies the cycle proceeds via reactions with relic photons.

At even lower particle energies quanta propagate without pair production and electrons scatter comparatively soft photons with mean energies (9.12). That results in fewer particles being produced in cascades. Therefore high energies are more favorable for cascade development.

9.4. THE MODEL ASSUMPTIONS

Author assumes that cosmic ray sources are active galactic nuclei of two types: BL Lac's with red shifts up to $z=1.1$ and Seyfert nuclei with red shifts $z \leq 0.0092$, and with space distributions according to red shift data from the catalogue by Veron-Cetty and Veron (2003).

In both cases author supposes that initial CR spectra are

1) monoenergetic with the energy of $E \sim 10^{21}$ eV or
2) exponential with the values of exponent of 2, 2.6, and 3.

Reasons for these were discussed in Section 7.

Author examines two options: (I) background radio emission energy and density are described following measurement results by Clark et al. (1982) and (II) when they are given by theoretical values (Protheroe and Biermann, 1997).

Author supposes that extragalactic magnetic field is $B \approx 10^{-11}$ G, so that electrons have negligible energy losses due to synchrotron emission even at energies of $E \approx 10^{20}$ eV. In the Galaxy, I do not consider ultra high energy electrons because they loose energy relatively fast in the magnetic field.

9.5. METHOD OF CALCULATIONS

Calculations based on the model above were performed for mean values with no fluctuations taken into account.

Author considers sources at distances farer than 500 Mpc assuming that particles from nearer sources do not initiate cascades. Also author ignores the microwave background evolution and the Universe inflation.

Author assumes also that in the cascade each quantum at energies $E_\gamma < 10^{15}$ eV reduces its energy to the value $E_\gamma = 10^{14}$ eV. However, the energy can be decreased to lower values, e.g. to $5 \cdot 10^{13}$ eV. Quanta at energies lower than 10^{14} eV loose energy interacting with optical photons and therefore the number N_γ can be less than estimates.

***

The proton free path was calculated using Monte-Carlo technique. Pion energy $E_{\pi 0}$ equals $E_{\pi 0}=KE_p$, where E_p is proton energy and the inelasticity coefficient K depends on the proton energy.

In decays (9.3), energy of gamma quanta is $E_\gamma=0.5E_{\pi 0}$.

Decays (9.5) produce positrons with energies in the range $1.3 \cdot 10^{-5} KE_p \leq E_e \leq KE_p$. For these calculations author assumes that the positron energy is small and positron contribution to the cascade is negligible. Author assumes that channels of proton interactions (9.1) and (9.2) are equally probable and considers only process (9.1). Then one half of interactions (9.1) give rise to the cascade.

For interactions described by (9.6), angle ψ is simulated in the observer system and the photon energy is calculated in the center-of- mass system. After that author determines cross section $\sigma_{\gamma\gamma}$ and the mean free path $\lambda_{\gamma\gamma}=1/(n_b\sigma_{\gamma\gamma})$ for radio photon with energy ε_b for $E_\gamma>E_t$ and for relic photons. Next step includes calculation of energy E_t, for the case of photon having the minimal path $\lambda_{\gamma\gamma}$.

Author assumes that in electron-positron pairs one particle has fixed energy E_t, while the second one - $(E_\gamma-E_t)$.

Electron interactions (9.7) are as follows. The cross section σ_{IC} of inverse Compton scattering for the case $E_e >E_t$ is calculated by formula (9.11). At lower energies the cross section equals σ_T. Then author calculates the mean free path $\lambda_{IC}=1/(n_b\sigma_{IC})$ assuming the radio photons or relic photons. Further author chooses the photon with the minimal path λ_{IC} and calculates the energy E_t. In the case $E_e <E_t$, the scattering cross section is σ_T, the scattered photon energy $E_{\gamma'}$ is calculated by (9.12), and the electron energy is $(E_e-E_{\gamma'})$. In the opposite case $E_e >E_t$, the scattered photon energy equals (E_e-E_t), the electron energy is E_t. The procedure is repeated until the electron energy decreases to 10^{14} eV.

9.6. CASCADES CHARACTERISTICS AFFECTED BY PARAMETERS OF RADIO BACKGROUND EMISSION

What cascade characteristics are most affected by radio background models?

In space, protons interact with background photons and initiate cascades at each point of interaction. Mean free paths of protons for several consequent interactions are shown in. Figure 8. For monoenergetic initial spectrum, protons have start energy $E=10^{21}$ eV. Protons loose a significant part of their energy to the cascade at distances not far from the source, $R\leq50$ Mpc. For the exponential initial spectrum major part of protons has energies $E<10^{20}$ eV and path values reach ~100 Mpc. So in this case energy transfer occurs relatively far from the sources.

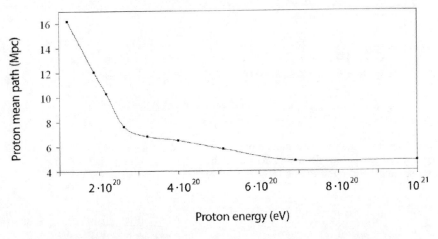

Figure 8. The proton mean path calculated as a function of the proton energy in interactions with relic photons, the initial proton energy being of 10^{20} eV.

Mean paths of quantum participating in the pair production at different energies are shown in figure 9. Two important conclusions are evident: first, mean paths do not exceed 10 Mpc for all initial energies; second, calculated curves depends on the model of radio background emission and differ considerably at quanta energies $E\approx10^{20}$ eV.

In inverse Compton scattering, electrons scatter background radio photons at energies $E>10^{19}$ eV. After electron energy decreases to $E\approx10^{19}$ - $5\cdot10^{18}$ eV electrons start scattering on relic photons. Therefore in the model (I) electrons transfer a significant part of their energy to radio photons and further to relic photons and do not scatter quanta with relatively low energies (9.12). In the model (II) electrons scatter radio photons with energies (9.12) until the energy decreases to approximately $5\cdot10^{18}$ eV, and after that electron scatters relic photons with energies $E_\gamma\approx E_e$. As a result in the model (I) the energy dissipation is considerably lower as compared to the model (II). Calculated electron mean paths are shown in figure 10.

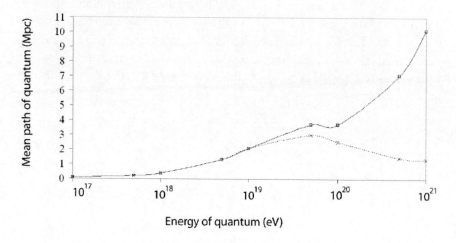

Figure 9. The calculated mean path of a quantum against the energy in the process of pair creating. Crosses denote values got in the radio background variant (I), squares show values got in the variant (II).

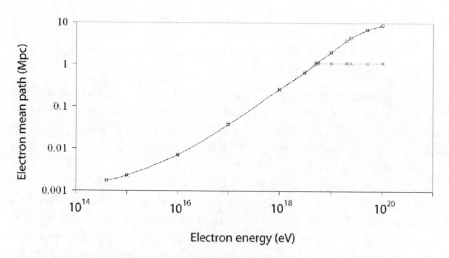

Figure 10. The electron mean path calculated as a function of the electron energy in the inverse Compton scattering. Crosses denote values got in the radio background variant (I), squares show values got in the variant (II).

9.7. RESULTS. NUMBER OF GAMMA QUANTA AT $E_\gamma \approx 10^{14}$ EV FOR DIFFERENT MODELS

Author examines two classes of sources: "distant" sources (BL Lac's) located at hundreds of Mpc away and "nearby" sources (Seyfert nuclei) located at tens of Mpc away, and two types of initial spectra in sources: the monoenergetic and the power-law. Reasons for these selections are discussed in Section 7.

First, author considers the "distant source" model, according to which BL Lac's emit protons with the energy $E_0 = 10^{21}$ eV. These protons interact with relic photons and the number of interactions is about 8. Each interaction generates a cascade which produces quanta. So the further the source more quanta can be generated. Let's consider cascade originated in the first proton-photon interaction (9.1). What is the number N_γ of quanta created in this cascade?

For sources at distances ~1000 Mpc, the number of quanta produced in the first cascade is $N_\gamma \sim 10^4$ for radio background (I), and $N_\gamma \sim 10^5$ for radio background (II). For sources at distances ~100 Mpc, the number of quanta is $N_\gamma < 10^3$ и $N_\gamma < 10^4$ respectively.

Now author considers sources with the exponential initial spectrum. In these models the number of produced quanta is considerably lower. This can be explained in the following way. Most of protons are emitted at energies $E \approx 5 \cdot 10^{19}$ eV. At this energy proton free paths are long; hence they interact with relic photons only once or twice and can initiate only one or two cascades. In addition, at lower energies fewer particles are originated in a cascade. As a result values of N_γ are smaller compared to N_γ when the primary proton energy is $E_0 = 10^{21}$ eV.

For sources at distances ~1000 Mpc, the cascade originated in the first interaction point produces $N_\gamma \sim 10^3$ and $N_\gamma \sim 10^4$ quanta for radio backgrounds (I) and (II), respectively. For sources at approximately 100 Mpc away, values of N_γ are negligible – cascades do not develop because the distance is too short.

And what are values of N_γ accounting for all proton-photon interactions and the source space distributions?

For distant sources emitting cosmic rays with the initial monoenergetic spectrum, number of quanta with energy $E_\gamma \approx 10^{14}$ eV near Earth equals to $N_\gamma \approx 3.3 \cdot 10^4$ with the radio background (I) and to $N_\gamma \approx 10^6$ with the radio background (II).

For distant sources emitting cosmic rays with the initial exponential spectrum number of quanta near Earth is $N_\gamma \approx 1.5 \cdot 10^3$ regardless of the radio background model.

Finally "nearby" Seyfert nuclei emitting protons with the initial power-law spectrum produces $N_\gamma \approx 0$ quanta near Earth.

9.8. Intensity of Quanta at $E_\gamma \approx 10^{14}$ eV in Cascades

Most cosmic rays are protons and atomic nuclei or nuclear fragments. It is a difficult to select showers produced by particles from those initiated by gamma rays. Nevertheless showers initiated by gamma rays were investigated and obtained results can be used in cosmic ray analysis.

At the energy $E_\gamma \approx 10^{14}$ eV, fraction of showers initiated by gamma ray compared to the total number of showers is

$$f = I_\gamma(10^{14} \text{ eV})/I_{CR}(10^{14} \text{eV}) = N_\gamma I_{CR}(E_0, \text{ eV})/ I_{CR}(10^{14} \text{ eV}), \qquad (9.19)$$

where I_γ is the intensity of gamma ray initiated showers, E_0 is the energy of protons which initiated cascades in the space, I_{CR} is cosmic ray intensity.

What are the intensities $I_{CR}(E_0, \text{ eV})$ and $I_{CR}(10^{14}\text{eV})$? To calculate that author uses the approximation of the cosmic rays integral intensity (it is applicable to use integral intensity because the cosmic ray spectrum decreases sharply). Cosmic ray integral intensity is given by expressions (Berezinsky et al. 1990):

$$I_{CR}(>E) = 1 \cdot (E \text{ (GeV)})^{-1.7} \text{ particle/(cм}^2 \cdot \text{s} \cdot \text{sr)}$$
$$\text{at } 10 \text{ GeV} < E < 3 \cdot 10^6 \text{ GeV ;} \qquad (9.20)$$

$$I_{CR}(>E) = 3 \cdot 10^{-10} (E \text{ (GeV)}/10^6)^{-2.1} \text{ particle/(cм}^2 \cdot \text{s} \cdot \text{sr)}$$
$$\text{at } E > 3 \cdot 10^6 \text{ GeV.} \qquad (9.21)$$

Author gets the following part f of showers which are produced by quanta at the energy $E_\gamma \approx 10^{14}$ eV.

For "distant" sources emitting protons with the initial monoenergetic spectrum, the fractions of gamma ray initiated showers at energy $E=10^{14}$ eV are $f{\approx}8{\cdot}10^{-7}$ and $f{\approx}3{\cdot}10^{-5}$ for radio backgrounds (I) and (II) respectively.

"Distant" sources emitting protons with the power-law spectrum produce f $\approx1.3{\cdot}10^{-13}$, for either radio backgrounds.

Finally, nearby Seyfert nuclei do not produce any gamma ray initiated showers at the energy 10^{14} eV.

Therefore, for BL Lac's, model with the initial monoenergetic spectrum predicts that for energy $E=10^{14}$ eV fractions of gamma ray initiated showers differs in ~40 times ($f{\approx}8{\cdot}10^{-7}$ vs. $f{\approx}3{\cdot}10^{-5}$), for two types of backgrounds. Next author compares these estimates with experimental data.

9.9. EXPERIMENT TO STUDY RADIO BACKGROUND AND INITIAL COSMIC RAY SPECTRA

Nikolsky et al. (1987) investigated showers initiated by gamma rays registered at the Tian-Shan array. Tian-Shan shower statistics at the energy $E=10^{14}$ eV is $N_{shower}{\approx}10^{8}$.

Author uses these data to suggest an experiment to study extragalactic radio background.

Let's consider BL Lac's with the initial monoenergetic CR spectrum. For the radio background of type (I) the number of gamma-ray initiated showers is $N_{(I)}=80$, while for type (II) this number is $N_{(II)}=3{\cdot}10^{3}$. The variation between these two numbers is big and this can be used for experimental studies of extragalactic radio background.

For the initial power-law CR spectrum in BL Lac's, the number of gamma-ray initiated showers is $N=0$, for shower statistics ~10^{8}. This result shows that both BL Lac's and Seyfert nuclei accelerate particles with power-law spectrum. The information about initial cosmic ray spectra is important because the spectrum shape depends on processes in which particles are accelerated in sources.

10. CONCLUSION

Ultra high energy cosmic rays are particles with energies $E > 4 \cdot 10^{19}$ eV. The most energetic particles which were registered had energies of $\sim 10^{20}$ eV and higher. Particle with the highest energy of $E \sim 3 \cdot 10^{20}$ eV was detected at the Fly's Eye array in the USA.

Cosmic ray investigations at ultra high energies are facing serious challenges related to low particles intensity, indirect methods of determining particle energies, low energy and angular resolution of arrays, compared to the methods used in astronomy.

Many of these problems have been successfully resolved by scientists from different countries all over the world. In this book author presented models that allow to make few steps towards better understanding of some of these problems.

The fundamental question in cosmic ray research is: how can we identify cosmic rays sources using data collected on the Earth? Many very interesting hypotheses were suggested related to this issue. Following other scientists author believes that these sources are extragalactic astrophysical objects. Hence it is possible to identify these objects. Author assumes that list of possible sources includes: active galactic nuclei, specifically "nearby" Seyfert galaxies at distances of 40-50 Mpc away, BL Lac's, radiogalaxies, as well as roentgen pulsars – the most powerful among pulsars. Using assumption that cosmic particles are only weakly deviated in the extragalactic magnetic field author made estimates of the strength of the field satisfying this assumption.

Author performed identification of sources analyzing shower arrival coordinates and calculating probabilities of random occurrence of objects in the error-boxes around the shower axis. It turned out that "nearby" low-luminosity Seyfert nuclei and BL Lac's appeared in the error-boxes around particle coordinates not randomly.

Presently scientists at Pierre Auger collaboration support the viewpoint that possible cosmic ray sources are not so distant low-luminosity Seyfert nuclei (The Pierre Auger Collaboration, 2007).

Author explained the existence of clusters of particles registered at AGASA in case when cosmic rays are generated by active galactic nuclei. According to astronomic observations variations in source activity result in cosmic ray sources switching on and off with breaks in activity being as long as approximately 25 years. So to register a cluster in addition to large area arrays longer observation periods are necessary.

Power necessary to accelerate cosmic rays in sources is discussed by (Hillas, 1984) and Aharonian et al. (2002). Active galactic nuclei satisfy limits in these papers. Active galactic nuclei are capable to generate particles to ultra high energies.

Author analyzed registered cosmic ray spectra assuming that particles are emitted by these two types of sources with initial spectra assumed to be either monoenergetic or power-law. It appeared that in these simplified cases it is possible to determine the shape of the initial spectrum using measured spectra. In addition, author concluded that in sources the maximal energy of accelerated particles is not higher than $E=10^{21}$ eV. Prior to that, this limit was obtained only theoretically.

Author suggested the model for cosmic ray acceleration in low-luminosity Seyfert nuclei. This model predicts chemical composition of cosmic rays. Estimates showed that ultra high energy rays are atomic nuclei or fragments with the charge $Z \geq 2$ rather than protons. Hence the model can be tested in the experiments related to cosmic ray chemical composition. In addition author suggested that it is possible to estimate magnetic fields in Seyfert nucleus jets using cosmic ray composition.

Finally author proposed to investigate gamma ray initiated showers with energy $E \approx 10^{14}$ eV. Some of gamma quanta are produced in extragalactic electromagnetic cascades. Cascades are originated when ultra high energy particles travel in the space. Selecting these quanta it is possible to study initial energy cosmic ray spectra in sources and also the extragalactic radio background. The energy $E \approx 10^{14}$ eV is chosen because the Universe is practically opaque for these quanta (Gould and Schreder, 1967), so quantum absorption is minimal.

10.1. PREDICTIONS OF DIFFERENT MODELS OF COSMIC RAY ORIGIN

A number of models explaining the origin of cosmic rays at ultra high energies have been suggested by different authors. Which predictions of these models can be tested in observations and cosmic ray experiments?

The model by (Haswell et al. 1992) predicts sporadic flares of radiation associated with ejections of accelerated particles. Model suggested by (Kardashev, 1995) also predicts certain characteristics of radiation from galactic nucleus in roentgen and gamma-ray bands.

In models for topological defects, the majority of cosmic particles at energies $E \approx 10^{21}$ eV should be gamma rays. However, it is not supported by AGASA results. Shinozaki et al. (2003) concluded that these are not quanta, but particles.

The model of gamma-ray bursts predicts that only ultra high energy protons are produced.

Kuz'min and Rubakov (1998) suggested the model in which cosmic rays at ultra high energies are produced in decays of relic particles generated in cold dark matter. If this is true there an appreciable (about 20%) excess of ultra high energy cosmic rays should be measured from the Galactic center. However no excess is observed experimentally.

Model discussed in this book predicts:

1. No cosmic ray excess from the Galactic center,
2. Ultra high energy cosmic rays from Seyfert nuclei are either atomic nuclei or nuclear fragments with Z>2,
3. Detected protons are not emitted by Seyfert nuclei, they are either nuclear fragments or are accelerated in BL Lac's.

Details of presented investigation can be found in (Uryson, 1996-2007).

ACKNOWLEDGMENTS

Author would like to thank A.V. Zasov for discussions of various astrophysical problems and N.S. Kardashev for discussions related to conditions in BL Lac's.

Author especially wants to thank B. Uryson who edited the manuscript of this book.

APPENDIX

Table. Equatorial coordinates (α, δ) and galactic latitudes b of shower arrival directions along with coordinates of the active galactic nuclei near to the shower axis. If an active galactic nucleus occurs in the region smaller than 2- or 3-error boxes the size of the region is given in units of Gaussian parameter σ. Showers #1-58 were detected by AGASA+A20, showers #59-62 were registered at Yakutsk array, and the shower #63 was detected at Haverah Park array.

	Ultra high energy showers				Seyfert nuclei			BL Lac's		
#	l	b	α	δ	α	δ		α	δ	
1	93.3°	-15.7°	22h21m	38.4°	22h08m	31.3°	2σ	22h50m	38°24'	2.4σ
2	63.5	19.4	18 29	35.3	-	-	-	18 13	31 44	1.25σ
3	170.4	-11.2	04 38	30.1	-	-	-	04 33	29 5	1σ
4	96.8	63.4	14 02	49.9	14 13	47.6	1σ	14 15	48 30	1σ
5	82.1	-21.1	21 57	27.6	22 08	31.3	2σ	-	-	-
6	68.3	75.6	13 48	34.7	13 45	35.6	1σ	13 40	27 43	1.7σ
7	154.5	15.6	05 51	58.5	-	-	-	05 58	53 28	1.7σ
8	6.1	29.6	16 17	-7.2	-	-	-	-	-	-
9	65.7	51.5	15 47	41.0	15 26	41.7	1.75σ	15 39	41 43	1σ
10	77.9	18.4	18 59	47.8	-	-	-	18 38	48 2	1.75σ
11	136.6	11.2	03 37	69.5	-	-	-	-	-	-
12	108.8	25.6	19 06	77.2	-	-	-	18 53	67 13	3.3σ
13	121.0	15.9	00 12	78.6	-	-	-	-	-	-
14	184.3	48.0	09 36	38.6	09 14	40.0	1.8σ	09 30	39 33	1σ
15	74.8	29.4	17 52	47.9	-	-	-	17 50	47 0	1σ
16	117.2	-45.0	00 34	17.7	-	-	-	00 35	15 15	1σ
17	53.6	35.6	17 03	31.4	-	-	-	17 02	31 15	1σ
18	143.2	56.6	11 29	57.1	11 22	59.0	1σ	09 30	39 33	1σ
19	180.5	13.9	06 44	34.9	-	-	-	06 43	42 14	2.4σ
20	206.7	26.4	08 17	16.8	-	-	-	08 29	17 54	1σ
21	108.8	55.5	13 55	59.8	13 41	67.7	3σ	13 53	56 0	1.3σ
22	139.8	-31.7	01 56	29.0	-	-	-	02 0	27 12	1σ
23	127.0	-12.7	01 16	50.0	00 42	41.3	3σ	01 23	42 10	2.6σ
24	130.5	-41.4	01 15	21.1	01 43	13.6	2.5σ	01 09	18 16	1σ
25	77.6	20.9	18 45	48.3	-	-	-	18 38	48 2	1σ
26	182.8	-15.5	04 56	18.0	-	-	-	05 02	13 38	1.5σ

#	l	b	α	δ	α	δ		α	δ	
27	145.5	55.1	11 14	57.6	11 22	59.0	1σ	10 58	56 28	1.3σ
28	22.8	15.7	17 37	-1.6	-	-	-	-	-	-
29	117.5	86.5	12 52	30.6	12 36	26.0	1.5σ	12 37	30 20	1.25σ
30	130.2	-42.5	01 14	20.0	-	-	-	01 09	18 16	1σ
31	171.1	-10.8	04 41	29.9	-	-	-	04 33	29 5	1σ
32	38.9	45.8	16 06	23.0	15 34	15.2	3σ	16 18	21 59	1σ
33	165.4	-20.4	03 52	27.1	-	-	-	03 22	23 36	2.5σ
34	105.1	29.8	17 56	74.1	-	-	-	17 32	69 26	2σ
35	113.8	63.7	13 18	52.9	13 29	47.2	2σ	13 24	57 39	1.6σ
36	56.8	-4.8	19 54	18.7	-	-	-	19 31	9 31	3σ
37	62.7	31.3	21 37	8.1	-	-	-	22 09	10 31	2.7σ
38	56.2	42.8	16 31	34.6	13 55	40.4	1.8σ	16 26	35 13	1σ
39	68.5	69.1	14 17	37.7	-	-	-	14 14	34 30	1.07σ
40	103.0	21.9	19 37	71.1	-	-	-	19 59	65 8	2σ
41	33.1	-13.1	19 38	-5.8	19 48	-10.3	1.5σ	20 3	-8 56	2.08σ
42	152.9	-43.9	02 18	13.8	02 25	18.5	1.6σ	02 17	8 37	1.7σ
43	171.2	64.6	11 09	41.8	11 05	46.4	1.5σ	11 0	40 19	1σ
44	207.2	48.6	09 47	23.7	09 19	26.3	2.2σ	09 40	26 3	1σ
45	84.5	35.3	17 16	56.3	-	-	-	17 05	60 42	1.5σ
46	147.5	56.2	11 13	56.0	11 22	59.0	1σ	10 58	56 28	1.25σ
47	89.5	-44.3	23 16	12.3	23 05	12.3	1σ	23 19	16 11	1.3σ
48	39.1	47.8	15 58	23.7	15 34	15.2	2.8σ	16 18	21 59	1.7σ
49	83.1	14.0	19 36	50.7	-	-	-	-	-	-
50	152.4	-7.8	03 45	44.9	-	-	-	03 13	41 15	2.6σ
51	39.9	-2.1	19 11	5.3	18 52	11.9	2.1σ	19 31	9 37	1.7σ
52	187.5	23.6	07 39	32.2	-	-	-	07 39	33 7	1σ
53	149.8	-4.0	03 46	49.5	-	-	-	03 13	41 15	2.75σ
54	98.5	-23.8	23 03	33.9	-	-	-	23 04	37 5	1σ

(Continued).

#	Ultra high energy showers				Seyfert nuclei			BL Lac's		
	l	b	α	δ	α	δ		α	δ	
55	98.8	-14.0	22 40	42.6	-	-	-	22 50	38 24	1.4σ
56	191.3	-26.5	04 37	5.1	-	-	-	04 22	2 19	1.25σ
57	150.3	-0.7	04 02	51.7	-	-	-	-	-	-
58	69.3	71.0	14 08	37.1	13 55	40.4	1.25σ	14 14	34 30	1σ
59	154.9	56.8	10 55	52.9	11 22	59.0	2.25σ	10 47	54 37	1σ
60	69.5	3.3	19 51	33.5	-	-	-	-	-	-
61	139.1	69.3	12 16	47.0	12 18	47.3	1σ	12 02	44 44	1σ
62	129.6	-16.4	01 25	45.7	01 09	35.7	3.3σ	01 23	42 10	1σ
63		78.4	11 56	27.0	12 15	33.2	2.1σ	11 49	24 38	1σ

REFERENCES

Afanasiev, B. N. et al. In International Symposium on Extremely High Energy Cosmic Rays: Astrophysics and Future Observatories; Nagano, M.; Ed.; Institute for Cosmic Ray Research. Tokyo, 1996; p. 32.

Aharonian, F. A. et al. *Phys. Rev.* 2002, D 66, 023005.

Aкerlof, C.W. et al. *Astrophys. J.* 2003. 586, 1232.

Antonucci, R. *Ann. Rev. Astron. Astrophys.* 1993, 31, 473.

Artyukh, V.S. et al. *Astron. Astrophys.* 2008, 486, 735.

Artyukh V.S. et al. *Astron. Rep.* 2009, 53, 988.

Beck, R. et al. *Ann. Rev. Astron. Astrophys.* 1996, 34, 155.

Bednarek, W. *Astrophys. J.* 1993, 402, L29.

Bednarek, W; Kirk, J. G. *Astron. Astrophys.* 1995, 294, 366.

Begelman, M.C. et al. *Rev. Mod. Phys.* 1984, 56, 255.

Berezinsky, V. S.; Grigor'eva, S. I. *JETP.* 1988, 93, 812.

Berezinsky, V.S. et al. *Astrophysics of Cosmic Rays;* Ginzburg, V.L; Ed.; Elsevier Science Publishers B.V: North-Holland, 1990.

Bird, D. et al. *Astrophys. J.* 1995, 441, 144.

Bhattacharjee, P.; Sigl, G. *Phys. Rep.* 2000, 327, 109.

Blandford, R.; Eichler, D. *Phys. Rep.* 1987, 154, 1.

Cesarsky, C.J. *Nucl. Phys. B* (Proc. Suppl.) 1992, 28, 51.

Chakrabarti, S.K. *Mon. Not. R. Astron. Soc.* 1988, 235, 33.

Clark, T.A. et al. *Nature.* 1970, 228, 847.

Cronin J.W. In International Symposium on Extremely High Energy Cosmic Rays: Astrophysics and Future Observatories; Nagano, M.; Ed.; Institute for Cosmic Ray Research. Tokyo, 1996; p. 2.

Dolag, K. et al. *JETP Lett.* 2004, 79, 583.

Dubovsky, S. L. et al. *Phys. Rev. Lett.* 2000, 85, 1154.

Falcke, H. et al. *Astron. Astrophys.* 1995, 298, 395.

Falcke, H. et al. *Astrophys. J.* 2000, 542, 197.

Field, G.B.; Rogers, R.D. *Astrophys. J.* 1993, 403, 94.

Gabuzda, D.C.; Cawthorne, T.V. *Mon. Not. R. Astron. Soc.* 2003, 338, 312.

Ginzburg, V.L.; Syrovatskii, S.I. The Origin of Cosmic Ray; MacMillan: New York, 1964.

Gould, R.J.; Schreder, G.P. *Phys. Rev.* 1967, 155, 1408.

Greisen, K. *Phys. Rev. Lett.* 1966, 16, 748.

Haswell, C.A. et al. *Astrophys. J.* 1992, 401, 495.

Hayakawa, S. Cosmic Ray Physics; John Willey & sons: New-York, 1969.

Hayakawa, S. *Progr. Theor. Phys. Suppl.* 1966, 37, 594.

Hayashida, N. et al., astro-ph /0008102, 2000.

Hillas A.M. *Ann. Rev. Astron. Astrophys.* 1984, 22, 425.

Hill, G.T.; Schramm D.N. *Phys. Rev.* 1985, D31, 564.

HiRes Collaboration. (2007). First Observation of the Greisen-Zatsepin-Kuzmin Suppression. arXiv:astro-ph/0703099.

Istomin, Ya.N.; Pariev, V.I. Mon. Not. R. *Astron. Soc.* 1994, 267, 629.

Kardashev, N.S. Mon. Not. R. *Astron. Soc.* 1995, 276, 515.

Krolik, J.H. *Astrophys. J.* 1999, 515, L73.

Kronberg, P.P. *Rep. Progr. Physics.* 1994, 57, 325.

Krymskii, F. *Dokl Akad. Nauk SS.* 1977, 234, 1306.

Kuhr, H. et al. *Astron. and Astrophys. Suppl. Ser.* 1981, 45, 367.

Landau, L.D.; Lifshits, E.M. Field Theory; Nauka: Moscow, 1990 [in Russian].

Lipovetsky, V.A. et al. Communications of the SAO; No 55; SAO AN SS, 1987.

Longair, M.S. High Energy Astrophysics; Cambridge University Press: Cambridge, 1981.

Medvedev, M. V. *Phys. Rev.* 2003, E67, 045401.

Nikolsky, S.I. et al. *J. Phys. G: Nucl. Phys.* 1987, 13, 883.

Norman, C.A. et al. *Astrophys. J.* 1995, 454, 60.

Parker, E. N. Proc. 22nd International Cosmic ray Conference. Dublin. 1991, 5, 35.

Particle Data Group, *Phys. Rev.* 2002, D69, 269.

The Pierre Auger Collaboration. *Science.* 2007a, 318, 938.

The Pierre Auger Collaboration. (2007b). [The UHECR spectrum measured at the Pierre Auger Observatory and its astrophysical implications.] arXiv:0707.2638 [astro-ph]

Popov, S.B. (2000) http://xrai.sai.msu.su/~polar/.

Prilutsky, O. F.; Rozental I. L. *Acta Phys. Hung. Suppl.* I. 1970, 29, 51.

Protheroe, R.J.; Biermann, P.L. *Astropart. Phys.* 1996, 6. 45; erratum, ibid. 1997, 7, 181.

Puget, J. L. et al. *Astrophys. J.* 1976, 205, 638.

Pyatunina, T.B. et al. Mon. Not. R. *Astron. Soc.* 2007, 381, 797.

Rees, M.J. 1987, Mon. Not. R. *Astron. Soc.* 1987, 228, 47.

Shatskii, A.A.; Kardashev N.S. *Aston. Rep.* 2002, 46, 639.

Shinozaki, K. et al. Proc. 28th International Cosmic ray Conference.Tsukuba. 2003, p. 401.

Sikora, M. et al. *Astrophys. J.* 1997, 484, 108.

Spinrad, H. et al. *Publ. Astron. Soc. Pac.* 1985, 97, 932.

Stecker, F. W. *Phys. Rev. Lett.* 1968, 21, 1016.

Stecker, F.W. et al. *Phys.Rev. Lett.* 1991, 66, 2697.

Stecker, F. W. *Phys. Rev. Lett.* 1998, 80, 1816.

Takeda, M. et al. *Astrophys. J.* 1999, 522, 225.

Tinyakov, P.G;, Tkachev, I.I. *JETP Lett.* 2001, 74, 445.

Thean, A. et al. Mon. Not. R. *Astron. Soc.* 2000, 314, 573.

Totani, T. *Astrophys. J. Lett.* 1998, 502, L13.

Ulvestad, J.S.; Ho L.C. *Astrophys. J.* 2002, 562, L133.

Uryson, A.V. *JETP Lett.* 1996, 64, 77.

Uryson, A.V. *JETP Lett.* 1997, 65, 729.

Uryson, A.V. *JETP.* 1999, 89, 597.

Uryson, A.V. *Astron. Astrophys. Transact.* 2001a, 20, 347.

Uryson A.V. *Astron. Rep.* 2001b, 45, 591.

Uryson, A.V. *Astron. Lett.* 2001c, 27, 775.

Uryson A.V. *Astron. Lett.* 2004, 30, 816.

Uryson A.V. *Astron. Lett.* 2005, 31, 755.

Uryson A.V. In Ultra High Energy Cosmic Rays: Possible Sources and Spectra. In *Frontiers in Cosmic Ray Research.* Martsch, I.N; Ed.; Nova Science: New-York, 2007; p 131.

Uryson A.V. *Bull. Russ. Acad. Sci. Physics.* 2007, 71, 914.

Veron-Cetty, M.P.; Veron P. *ESO Scientific Report.* N 13. 1993.

Veron-Cetty M.P.; Veron, P. *ESO Scientific Report.* N 19. 1998.

Veron-Cetty, M.P.; Veron P. *Astron. Astophys.* 2001, 374, 92.

Veron-Cetty, M.P.; Veron P. *Astron. Astophys.* 2003, 412, 399.

Vilenkin. A. *Nucl. Phys. B. Proc. Suppl.* 1996, 48, 508.

Vil'koviski,i E.Ya.; Karpova, O.G. *Astron. Lett.* 1996, 22, 168.

Wielebinski, R., Beck, R. *Cosmic Magnetic Fields;* Springer: Berlin, 2005.

Watson A., In Particle and Nuclear Astrophysics in the Next Millenium; Kolb, E.W.; Peccei R.D.; Ed.; World Scientific: Singapore, 1995; p. 126.

Watson, A. In International Symposium on Extremely High Energy Cosmic Rays: Astrophysics and Future Observatories; Nagano, M.; Ed.; Institute for Cosmic Ray Research. Tokyo, 1996; p. 362.

Wright, G.S. et al. *Mon.Not.R. Astron. Soc.* 1988, 233, 1.

Xu, C. et al. *Astron. J.* 1999, 118, 1169.

Zakharov, A.F. et al. *Mon. Not.R. Astron. Soc.* 2003, 342, 1325.

Zatsepin, G.T.; Kuz'min V.A. *JETP Lett.* 1966, 4, 78.

Zheleznyakov, V.V. 1997, Radiation in Astrophysical Plasma; Yanus-K: Moscow, 1997 [in Russian].

Zheleznyakov, V.V.; Koryagin, S.A. *Astron. Lett.* 2002, 28, 809.

INDEX